Catherinus Putnam Buckingham

Elements of the Differential and Integral Calculus,

By a New Method founded on the True System of Sir Isaac Newton, Without the use of Infinitesimals or Limits

Catherinus Putnam Buckingham

Elements of the Differential and Integral Calculus,
By a New Method founded on the True System of Sir Isaac Newton, Without the use of Infinitesimals or Limits

ISBN/EAN: 9783744729413

Printed in Europe, USA, Canada, Australia, Japan

Cover: Foto ©berggeist007 / pixelio.de

More available books at **www.hansebooks.com**

ELEMENTS

OF THE

DIFFERENTIAL AND INTEGRAL

CALCULUS,

BY A NEW METHOD, FOUNDED ON THE TRUE SYSTEM OF SIR ISAAC NEWTON, WITHOUT THE USE OF INFINITESIMALS OR LIMITS.

REVISED EDITION.

BY C. P. BUCKINGHAM,

AUTHOR OF THE PRINCIPLES OF ARITHMETIC; FORMERLY ASSISTANT PROFESSOR OF NATURAL PHILOSOPHY IN THE U. S. MILITARY ACADEMY, AND PROFESSOR OF MATHEMATICS AND NATURAL PHILOSOPHY IN KENYON COLLEGE, OHIO.

CHICAGO:
S. C. GRIGGS & COMPANY.
1880.

DONNELLEY, GASSETTE & LOYD, PRINTERS, CLARK & ADAMS STREETS.

PREFACE TO THE FIRST EDITION.

"The student of mathematics, on passing from the lower branches of the science to the infinitesimal analysis, finds himself in a strange and almost wholly foreign department of thought. He has not risen, by easy and gradual steps, from a lower to a higher, purer and more beautiful region of scientific truth. On the contrary, he is painfully impressed with the conviction, that the continuity of the science has been broken, and its unity destroyed, by the influx of principles which are as unintelligible as they are novel He finds himself surrounded by enigmas and obscurities, which only serve to perplex his understanding and darken his aspirations after knowledge."*

He finds himself required to ignore the principles and axioms that have hitherto guided his studies and sustained his convictions, and to receive in their stead a set of notions that are utterly repugnant to all his preconceived ideas of truth. When he is told that one quantity may be added to another without increasing it, or subtracted from another without diminishing it—that one quantity may be infinitely small, and another infinitely smaller, and another infinitely

*Bledsoe—Philosophy of Mathematics.

smaller still, and so on *ad infinitum*—that a quantity may be so small that it can not be divided, and yet may contain another an indefinite, and even an *infinite* number of times—that zero is not always nothing, but may not only be something or nothing as occasion may require, but may be *both* at the *same time* in the same equation—it is not surprising that he should become bewildered and disheartened. Nevertheless, if he study the text books that are considered orthodox in this country and in Europe he will find some of these notions set forth in them all; not. indeed, in their naked deformity, as they are here stated, but softened and made as palatable as possible by associating them with, or concealing them beneath, propositions that are undoubtedly true.

It is, indeed, strange that a science so exact in its results should have its principles interwoven with so much that is false and absurd in theory; especially as all these absurdities have been so often exposed, and charged against the claims of the calculus as a true science. It can be accounted for only by the influence of the great names that first adopted them, and the indisposition of mathematicians to depart from the simple ideas of the ancients in reference to the attributes of quantity. They regard it as merely inert, either fixed in value or subject only to such changes as may be arbitrarily imposed upon it. But when they attempt to carry this conception into the operations of the calculus, and to account for the results by some theory consistent with this idea of quantity they are inevitably entangled in some of the absurd notions that have been mentioned. Many efforts have indeed been made to escape such glaring inconsistencies, but they have only resulted in a partial success in concealing them.

To clear the way for a logical and rational consideration of the subject, we must begin with the fundamental idea of the conditions under which quantity may exist. We must, for the purposes of the calculus, consider it not only as *capable* of being increased or diminished, but also as being actually in a *state of change*. It must (so to speak) be *vitalized*, so that it shall be endowed with *tendencies* to change its value, and the rate and direction of these tendencies will be found to constitute the ground work of the whole system. The differential calculus is the SCIENCE OF RATES, and its peculiar subject is QUANTITY IN A STATE OF CHANGE.

It is an error, therefore, to suppose, as has often been said, that the "*reductio ad absurdum*," or "method of exhaustion," of the ancient mathematicians, contains the germ of the differential calculus. This hallucination has arisen from the same source as the false notions before mentioned. That peculiar attribute of quantity upon which the transcendental analysis was built, never found a place among the ideas of the Greek Philosophers; and even Leibnitz, the competitor of Newton for the honor of the invention, and who was the first to construct a system of rules for the *analytical machinery* of the science, never got beyond the ancient conception of the conditions of quantity, and, therefore, gave, says Comte, "an explanation entirely erroneous"—he never comprehended the true philosophical basis of his own system.

The only original birth-place of the fundamental idea of

quantity which forms the true germ of the calculus, was in the mind of the immortal Newton. Starting with this idea, the results of the calculus follow logically and directly through the beaten track of mathematical thought, with that clearness of evidence which has ever been the boast of mathematics, and which leaves neither doubt nor distrust in the mind of the student.

To develop this idea is the object of this work.

C. P. BUCKINGHAM.

CHICAGO, *Jan.* 1, 1875.

PREFACE TO THE PRESENT EDITION.

In a former edition of this work, I have condemned both the method of infinitesimals and the method of limits, as the true exponents of the Differential Calculus; the first, because its principles are manifestly false; and the latter, because no satisfactory demonstration had been given, that its assumptions were mathematically true.

Subsequent reflection, while it has not changed my opinion as to the weakness of the demonstrations or arguments usually advanced in support of the method of limits, has convinced me that by adopting a right conception of the true principles of the calculus, we may fathom the mystery of that method, and give it an honored place in the temple of scientific truth.

The method of limits has usually been regarded as resting upon the principle of exhaustion, and when demonstrated or defended on that ground, the attempt has always been a failure. It therefore seemed to me, that when such men as Carnot, Lagrange, D'Alembert, Maclaurin and even Newton himself, as well as a host of others, after a discussion during two hundred years, had failed to give a rational demonstration of the true meaning of $\frac{o}{o} = 2x$, as it arises in the calculus, it must be impossible to do it. Hence I took that ground, and, on it, objected to the method of limits as defective in its logic, and hence without a valid claim to a place in mathematics.

I have now, however, constructed a rigorous mathematical demonstration, entirely free from all metaphysical reasoning and which, I believe, is now set before the public, *for the first time*, showing that the equation $\frac{0}{0} = 2x$ is not only significant, but that $2x$ in that equation is the true differential coefficient, and that the hitherto unauthorized substitution of $\frac{dy}{dx}$ for $\frac{0}{0}$ is perfectly justifiable. This demonstration, which will be found in these pages, is based on the principle which is at the foundation of this work.

The practical *value* of this method, even when proved to be right in principle, must be estimated by considerations which will be set forth in another place.

C. P. BUCKINGHAM.

CHICAGO, *July* 1, 1880.

Contents.

INTRODUCTION.

PART I.

DIFFERENTIAL CALCULUS.

PART II.

INTEGRAL CALCULUS.

INTRODUCTION.

THE PHILOSOPHY OF THE CALCULUS.

In the early history of geometry, among a rude people, whose ideas were generally bounded by that, only, which could be seen and tested by experiment, the science was necessarily confined to the solution of the most simple propositions. After the measurement of rectangles, it was easy to proceed to that of parallelograms, and thence to triangles; and as all rectilinear figures could be divided into triangles, they were able to find the properties of a great variety of such figures without difficulty. Similarly, the measurement of rectangular parallelopipeds led to that of those which were oblique, and thence to the measurement of prisms with triangular and polygonal bases. But when they came to figures bounded by curved lines, and volumes bounded by curved surfaces, the primitive methods which they had used, were found to be powerless; and it was not until philosophy began to be cultivated, and men began to look beyond the mere evidence of their senses, and to cultivate ideas of abstract truth, that any important advance was made in the science. At length some unknown philosopher struck out a new method which opened up a brilliant career to geometry. The exact period of its invention is not known, but the first extensive use of it was made in the works of Euclid and Archimedes, who rapidly enlarged the boundaries of mathematical knowledge and brought the science of geometry to such a degree of perfection that it remained for two thousand years, without further progress. The method which, in their hands, worked such wonders is called,

THE METHOD OF EXHAUSTION.

It is also called the "*reductio ad absurdum*," because it shows that every supposition but the true one leads to an absurdity.

* "As the ancients" says Carnot "admitted only demonstrations which are perfectly rigorous, they believed they could not permit themselves to consider curves as polygons of a great number of sides; but when they wished to discover the properties of any one of them, they regarded it as a fixed term, which the inscribed and circumscribed polygons continually approached, as nearly as they pleased, in proportion as they augmented the number of their sides. In this way, they exhausted in some sort the space between the polygons and the curves; which, without doubt, caused to be given to this procedure the name of the *method of exhaustion*."

† "This will, perhaps be more clearly seen in an example. Suppose, then, that regular polygons of the same number of sides are inscribed in two circles of different sizes. Having established that the polygons are to each other as the squares of their homologous lines, they concluded, by the method of exhaustion, that the circles are to each other as the squares of their radii. That is, they supposed the number of the sides of the inscribed polygons to be doubled, and this process to be repeated until their peripheries approached as near as we please to the circumferences of the circles. As the spaces between the polygons and the circles were continually decreasing, it was seen to be gradually exhausted; and hence the name of the method. But although the polygons, by thus continuing to have the number of their sides doubled, might be made to approach the circumscribed circles more nearly than the imagination can conceive, leaving no appreciable difference between them; they would always be to each other as the squares of their homologous sides, or as the squares of the radii of the circumscribed circles. Hence they conjectured that the circles themselves, so very like the polygons in the last stage of their fullness or roundness, were

* Reflexions sur la Metaphysique du calcul infinitesimal.

† Bledsoe — Philosophy of Mathematics.

to each other in the same ratio, or as 'the squares of their radii.' But it was the object of the ancient geometers, not merely to divine, but to demonstrate. A perfect logical rigor constituted the very essence of their method. Nothing obscure, nothing vague, was admitted, either into their premises, or into the structure of their reasoning. Hence their demonstrations absolutely excluded the possibility of doubt or controversy; a character and a charm which, it is to be lamented, the mathematics has so often failed to preserve in the spotless splendor of its primitive purity.

"Having divined that any two circles (C and c) are to each other as the squares of their radii (R and r), the ancient geometers proceeded to demonstrate the truth of the proposition. They proved it to be necessarily true by demonstrating every other possible hypothesis to be false."

Thus, said they, the inscribed polygons P and p are to each other as $R^2 : r^2$, and if P is not to C as p : c then let us suppose P : C : : p : c′ ; c′ being a circle smaller than c. Increase the number of sides of the polygons, until p becomes greater than c′; (for we can make p approach c as near as we please) then we shall have P : C : : p : c′ in which the first consequent is greater than its antecedent, while the second consequent is less than its antecedent, which is absurd; and we can not have P : C : : p : a circle less than c. By a similar course of reasoning, using the circumscribed polygons, it may be shown that we can not have P : C : : p : a circle greater than c. Hence the proportion P : C : : p : c is true, and therefore P : p : : C : c : : $R^2 : r^2$. This process, by which every posssible supposition, except the one to be demonstrated was shown to lead to an absurdity, has also been called the "*reductio ad absurdum*," as well as the "*method of exhaustion.*"

* "By this method the ancients demonstrated all the intricate problems in elementary geometry, and brought that science to the condition in which it remained for two thousand years. Truths were waiting on all sides to be discovered, and continued to wait for centuries, until a more powerful instrument of discovery could be invented."

* Bledsoe — Philosophy of Mathematics.

At length the period of stagnation was ended, and in 1637 the brilliant genius of Descartes, seizing upon a new idea, boldly followed its lead until he developed a system whose results astonished and delighted the world. This system is that of

ANALYTICAL GEOMETRY.

Breaking away from the idea of determinate values and absolute conditions, he adopted that of dependent conditions and relative values, which no longer fixed unchangeably the quantities sought, but gave them a wide range, so that within certain limits they could have all possible values. Hence they were called *variables*, while those quantities whose values were fixed were called *constants*.

In every equation containing a single unknown quantity, the value of that quantity is absolutely fixed by the conditions expressed in the equation. If we have two unknown quantities, and two equations, or sets of conditions, both values are still fixed. If the higher power of the unknown quantity is involved, the number of values is greater, but they are equally fixed and certain. This idea of fixedness of value underlies all algebraic operations of an ordinary kind.

Now suppose we have *two* unknown quantities in *one* equation, with no other conditions given than those expressed by the equation itself. In that case the values of both quantities are absolutely indeterminate. But if we know or assume any specific value for one, we can at once determine the corresponding value of the other; so that while the equation will give the independent value of neither quantity, it will give the *simultaneous values of both;* and these values will have a certain range or locus, which is in fact the true solution of the equation — the path, so to speak, through which the simultaneous values range.

In some equations the range of values is limited for both variables, so that if a value be assigned to one beyond the limit, that of the other becomes imaginary; in other cases the value of one only is limited, while in others again the values of both are absolutely unlimited; any value of one giving a corresponding real value for the other.

Since the values of these variables are thus dependent on each other, the equation expressing this dependence may be considered as containing the *law* of their mutual relations, and the fundamental idea of Descartes was to exhibit in his equation the conditions or law which confined the two variables to their prescribed range of values. This idea was something new, distinct and well defined, and a clear advance beyond the methods of the ancients.

But the labors of Descartes would have been of little value, had he proceeded no farther than we have indicated. In fact this was but a part of his invention, of which the specific object was a method of investigating questions of Geometry. To complete this purpose, he devised a new and beautiful method of representing magnitudes, to which his algebraic equations could be applied. In algebra, all values are estimated by their remoteness from zero. In order to make a general application of algebraic symbols to geometry, it was necessary that the value of every line represented by his variables should be estimated from a common origin corresponding with zero; and as every point in a plane surface requires two values to fix its position, two such origins became necessary to his system, in order to represent plain figures; and these were found in two right lines, lying in the plane of the object to be represented, and intersecting in a known angle — generally a right angle. From these two lines all values or distances to points were estimated; the positive on one side and the negative on the other, of each line; while for points *in* the lines, one of the values would of course be zero

Having then a method of representing the position of a point by algebraic symbols, it was easy to apply his analysis to the representation of lines, by making the locus or range of simultaneous values of the variables to correspond with the locus of the points in the line — that is, with the line itself.

Thus the method of Descartes was two-fold — the algebraic idea of two variables in one equation with a range of simultaneous values, and the geometrical idea of coördinate representation, and these two being adapted to each other, united to form a method of investigating, in an easy and simple manner, questions of geometry which had taxed the utmost powers of the ancients.

The invention of Descartes did not, however, change the conceptions that had been formed of the nature and composition of quantities. A rectangle was still one entire surface, and a cone and sphere were still "solids."

The first important departure from the old ideas about magnitudes, was when Kepler introduced the notion that all magnitudes are made up of an infinite number of infinitely small parts. In 1635 Cavalieri modified this conception somewhat, and formulated it into a system, which he published under the title of the

METHOD OF INDIVISIBLES.*

"In this method, which opened a new era in Geometry, lines are conceived to be made up of points, surfaces composed of lines, and volumes composed of surfaces."

This idea was a startling one, and was not received without very decided objections. "These hypotheses," says Carnot, "are certainly absurd, and ought to be employed

* In discussing the method of indivisibles the writer has drawn largely on "The Philosophy of Mathematics," by Dr. A. T. Bledsoe.

with circumspection." One would naturally inquire, if they are absurd, why employ them at all? The reason, which seems to have reconciled Carnot to their use, is, that they produce true results. Hence, as if by way of apology, he says:

> "It is necessary to regard them as means of abbreviation, by means of which we obtain, promptly and easily, in many cases, what could be discovered only by long and painful processes, according to the method of exhaustion." "The great geometers, who followed this method, soon seized its spirit; it was in great vogue with them until the discovery of the new calculus, and they paid no more attention to objections that were then raised against it than the Bernouillis paid to those that were afterwards raised against the infinitesimal analysis. It was to this method of indivisibles that Pascal and Roberval owed their profound researches concerning the cycloid."

"Thus," says Dr. Bledsoe, "while appealing to the practical judgment of mankind, they treated the demands of our rational nature with disdain, and the more so, perhaps, because these demands were not altogether silent in their own breasts."

To assume the truth of a theory because it accounts for all the facts, can not be admitted in mathematics, because mathematical truth must admit of no doubt in the mind of him who comprehends it; and in this kind of reasoning there is always, at least, one source of doubt, namely, that there may be other theories that will, equally well, account for the same facts; which is indeed eminently the case in this very subject. Hence our first enquiry should be, "is the theory based on principles, the truth of which commends itself to every man's consciousness, and by which the theory can be proven true?" "Cavalieri acknowledged," says Carnot, "that he could not give a rigorous demonstration of his method." He asserted however that it was simply a corollary from the method of exhaustion — an assertion that is not very clear to an ordinary mind.

*" But Pascal himself, though universally recognized as one of the greatest geniuses that ever lived, could not comprehend the hypotheses or postulates of the method of indivisibles as laid down by Cavalieri. Hence, while he continued to use the language of Cavalieri, he attached a different meaning to it — a change which is supposed by writers on the history of mathematics to have improved the rational basis of the method. 'By an indefinite number of lines' said he 'he always meant an indefinite number of small rectangles,' 'of which the sum is certainly a plane.' In like manner, by the term 'surfaces' he meant 'indefinitely small solids,' the sum of which would surely make a solid. Thus he concludes, 'if we understand, in this sense, the expressions *the sum of the lines, the sum of the planes, etc.*, they have nothing in them but what is perfectly conformed to pure geometry." This is true. The sum of the planes is certainly a plane, and the sum of little solids is certainly a solid. But from this point of view it is surely not proper to call it 'the method of indivisibles,' since every plane as well as every solid may easily be conceived to be *divided.* The improved postulates of Pascal deliver us, indeed, from the chief difficulty of the method of indivisibles, properly so called, only to plunge us into another — into the very one, in fact, from which Cavalieri sought to effect an escape by the invention of his method."

" Thus, if we divide any curvilinear figure into rectangles, no matter how small, the sum of these rectangles will not be exactly equal to the area of the figure. On the contrary, this sum will differ from that area by a surface equal to the sum of all the little mixtilinear figures at the ends of the rectangles (Fig. 1). It is evident, however, that the smaller the rectangles are made, or the greater the number becomes, the less will be the difference in question. But how could Cavalieri imagine that this difference would ever become absolutely nothing, so long as the inscribed rectangles continue to be surfaces? Hence, in order to get rid of this difference altogether, and to arrive at the exact area of the proposed figure, he conceived the small rectangles to increase in number until they dwindled into veritable lines. The sum of these lines he supposed would be equal to the area of the figure in question, and he was confirmed in this hypothesis, because it was found to lead to perfectly exact results. Thus, his hypothesis was adopted by him, not because it had appeared at first, or in itself considered, as intuitively

Fig.1

*Bledsoe—Philosophy of Mathematics.

certain, but because it appeared to be the only means of escape from a false hypothesis, and because it led to so many exactly true results. But when this hypothesis, abstractly considered, was found to shock the reason of mankind, who, in the words of Carnot, pronounced it 'certainly absurd,' the advocates of the method of indivisibles were obliged to assume new ground. Accordingly, they discovered that indivisibles might be divided, and that by 'the sum of right lines' was only meant 'the sum of the indefinitely small rectangles.' Pascal seems to believe, in fact, that such was the meaning of Cavalieri himself.

"Now it is just as evident that a curvilinear figure is not composed of rectangles, as that it is not composed of right lines. Yet Pascal, the great disciple, adopted the supposition as the only apparent means of escape from the absurdity imputed to that of the master; and he pointed to the perfect accuracy of his conclusions as a proof of the truth of his hypothesis. For, strange to say, the sum of the rectangles, as well as the sum of the lines, was found to be exactly equal to the curvilinear figure. What, then, became of the little mixtilinear figures at the extremities of the rectangles? How, since they were omitted or thrown out, could the remaining portion of the surface, or the sum of the rectangles alone be equal to the whole? Pascal just cut the Gordian knot of this difficulty by declaring that if two finite quantities 'differ from each other by an indefinitely small quantity' then 'one may be taken for the other without making the slightest difference in the result.' Or, in other words, that an infinitely small quantity may be added to, or subtracted from a finite quantity without making the least change in its magnitude. It was on this principle 'that he neglected without scruple,' as Carnot says, 'these little quantities as compared with finite quantities; for we see that Pascal regarded as simple rectangles, the trapeziums or little portions of the area of the curve comprised between the two consecutive ordinates, neglecting consequently the little mixtilinear triangles which have for their bases the differences of those ordinates.'*

"Carnot adds, as if he intended to justify this procedure, that 'no person, however, has been tempted to reproach Pascal with want of severity.' This seems the more unaccountable, because Carnot himself has repeatedly said that it is an error to throw out such quantities as nothing. Nor is this all. No one can look the principle fairly and fully in the face, that an infinitely small quantity may be subtracted from a finite quantity without making even an infinitely small difference in its

*Carnot, Chap. III., p. 146.

value, and yet regard it as otherwise than absurd. It is only when such a principle is recommended to the mathematician by the desperate exigencies of a system which strains his reason, warps his judgment, and clouds his imagination, that it is admitted to a resting place in his mind. It was thus, as we have seen, that Pascal was led to adopt the principle in question; and it was thus, as we shall see, that Leibnitz was induced to assume the same absurd principle as an unquestionable axiom in geometry."

"Now if, with Cavalieri, we suppose a surface to be composed of lines; or a line of points, then we shall have to add points or no-magnitudes together until we make magnitudes. Nay, if lines are composed of points, surfaces of lines, and solids of surfaces, then it is perfectly evident that solids are made up of points, and the very largest magnitude is composed of that which has no magnitude! or, in other words, every magnitude is only the sum of nothing! On the other hand, if we agree with Pascal that a curvilinear space is, strictly speaking, composed of rectangles alone, then we shall have to conclude that one quantity may be taken from another without diminishing its value! Which term of the alternative shall we adopt? On which horn of the dilemma shall we choose to be impaled? But is it, indeed, absolutely necessary to be swamped amid the zeros of Cavalieri, or else to wear the yoke of Pascal's axiom? May we not by a recurrence to the true principles of mathematical philosophy find a safe passage between this Scylla and Charybdis of the infinitesimal method?"

We shall see further on.

The following beautiful example will illustrate most clearly and favorably the method of indivisibles. It is given by Carnot from Cavalieri and Pascal to recommend that method:

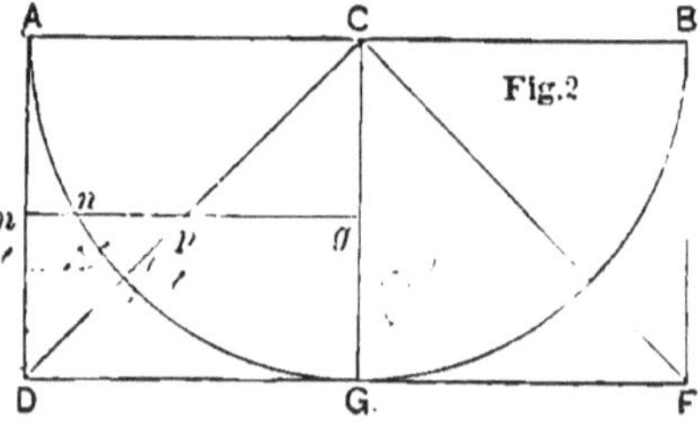

"Let A B (Fig 2)," says he, "be the diameter of a semi-circle, A G B; let A B F D be the circumscribed rectangle; C G the radius perpendicular to D F; let the two diagonals C D, C F also be drawn; and finally through any point *m* of the line A D, let the right line *m n p g* be drawn perpendicular to C G, cutting the circumference of the circle at the point *n*, and the diagonal C D at the point *p*."

"Conceive the whole figure to turn around C G as an axis; the quadrant of the circle A C G will generate the volume of a semi-sphere whose diameter is A B; the rectangle A D G C will generate the circumscribed right cylinder; the isosceles right-angled triangle C G D will generate a right cone, having the equal lines C G, D G for its height and for the radius of its base; and finally the three right lines or segments of a right line *m g, n g, p g*, will each generate a circle, of which the point *g* will be the centre.

"But the first of these circles is an element of the cylinder; the second is an element of the semi-sphere, and the third, of the cone.

"Moreover the areas of these circles are as the squares of their radii, and as these three radii can evidently form the hypothenuse and the two sides of a right angle triangle, it is clear that the first of these circles is equal to the sum of the other two; that is to say, that the element of the cylinder is equal to the sum of the corresponding elements of the cone and semi-sphere; and as it is the same with all the other elements cut out by one horizontal plane, it follows that the total volume of the cylinder is equal to the sum of the total volumes of the semi-sphere and cone. But we know that the volume of the cone is equal to one-third that of the cylinder; then that of the semi-sphere is two-thirds of the volume of the circumscribed cylinder, as Archimedes discovered."

We see here that we may, with equal propriety, with Cavalieri, consider the elements as mere surfaces, with no thickness, in which case we have something made up of nothing; or, with Pascal, as infinitely thin cylinders, in which case we must ignore the slender rings, surrounding these cylinders, and whose interior sides, in the one case, form the actual surface of the cone, and whose exterior sides, in the other case, form the surface of the semi-sphere.

The following is the true solution of this proposition: Suppose the cylinder, semi-sphere and cone to be generated, each by its element, commencing together at the line D F and flowing upwards at equal uniform rates, expanding, contracting, or remaining constant, so that each element shall at all times have its circumference in the surface of the body it is generating. The rate at which each volume is generated at any moment will be in proportion to the area of the generatrix at that moment; and since the sum of the generatrices of the semi-sphere and cone is, at all times, equal to the generatrix of the cylinder, the sum of the magnitudes generated by them will be equal to the magnitude of the cylinder.

Behold, then, the solution which avoids both horns of the dilemma.

"It is true that this axiom of Pascal has high authority in its favor. Roberval, Pascal, Leibnitz, the Marquis de L'Hopital, and others, have all lent the sanction of their great names to support it, and give it currency in the mathematical world. But does a real axiom ever need the support of authority?

On the other hand, there is against this pretended axiom, as intrinsically and evidently false, the authority of men equally, or more, celebrated in the mathematical world, such as Berkeley, Maclaùrin, Carnot, Euler, D'Alembert, Lagrange and Newton, whose names preclude the mention of any others. The very fact that mathematicians disagree proves that it is not about an axiom, but only about something else which has been set up as an axiom. It is indeed the very essence of geometrical axioms that they are necessary and universal truths, absolutely commanding the assent of all, and shining like stars above the dust and darkness of human controversy."

The next important movement in the mathematical world was when Newton and Leibnitz brought out their systems, in which the results are essentially the same, while the fundamental idea or philosophy of each is entirely different from the other.

METHOD OF LEIBNITZ.

In 1684 Leibnitz first published his Differential Calculus, in which he gave the first system of rules for the required operations. His method was formed on a conception similar to that of Pascal; viz. that all quantities were composed of infinitesimals. He considered surfaces bounded by curved lines as only polygons whose sides were infinitely small straight lines, and whose areas were composed of an infinite number of infinitesimal rectangles, while the tangent lines to curves were the prolongations of the infinitely small polygonal sides — that solids were composed of an infinite number of infinitely small cylinders or prisms, and that a double curved surface was in fact that of a polyedron with infinitely small faces, which, being extended would form tangent planes to the surface.

The whole superstructure of the method of Leibnitz rests upon two assumptions which are thus described by the Marquis de L'Hopital.

First. " We demand that we can take indifferently the one for the other, two quantities which differ from each other by an infinitely small quantity; or (what is the same thing), that a quantity which is increased or diminished by another quantity infinitely less than itself, can be considered as remaining the same.

Second. " We demand that a curved line can be considered as the assemblage of an infinity of right lines, each infinitely small; or (what is the same thing), as a polygon with an infinite number of sides, each infinitely small, which determine by the angles which they make with each other the curvature of the lines."

But Leibnitz carried his idea much further than Pascal, for he considered each of the infinitesimals of his system as also composed of an infinite number of parts, infinitely smaller than itself; and these, again, composed of an infinite number of parts, infinitely smaller yet: and so on indefinitely. He called them infinitesimals of the first, second, third, and so on, orders, in which each particle or portion of any order is infinitely less than one of the preceding order, and bore the same relation to it that the infinitesimal of the first order bore to the original function.

He indicated these infinitesimal increments, parts, or differences, by the term "differentials," and thus we find in his system, differentials of the first, second, third, and so on, orders; which he called, "first differential, second differential," and so on.

In order to illustrate the peculiarity of the idea of Leibnitz, let us suppose a curve to be composed of an infinite number of right lines, infinitely short, which, for the sake of perspicuity we will represent by the polygon A B C D E.

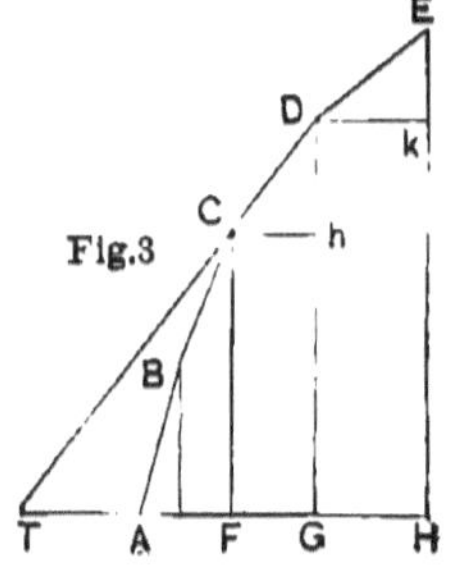

Let C F and D G (Fig. 3) be consecutive ordinates of the curve and C D an infinitesimal side lying between them. The infinitesimal difference D h between these ordinates is called the differential of the ordinate at the point C, and this, divided by the corresponding differential C h (or F G) of the abscissa, will give the tangent of the angle C T F which the side C D, or the tangent line C T, which is a prolongation of the side C D, makes with the axis of abscissas. But the position of this tangent line does not decide which way the curve is bending, that is, whether it is concave or convex toward the axis of abscissas. To ascertain this last point; the next infinitesimal side D E of the curve is taken, and the infinitesimal difference E k between its two ordinates is compared with the infinitesimal difference D h between the ordinates of the preceding side C D, and the difference between these two infinitesimal differences will show which way the curve is bending. If this difference is negative, the second infinitesimal E k is less than the first, and the curve is concave toward the axis of abscissas; if it is positive, the last infinitesimal difference is greater, and the curve is convex toward that axis.

The difference between these two infinitesimal differences is called the second differential of the ordinate; and as the first differential is infinitely less than the ordinate C F, so the second differential is infinitely less than the first.

Now this method of differences when the figure is actually a polygon is reasonable enough, but when applied to a real curve, represented by an *equation*, we find that the theory of the many sided polygon will not account for the results.

Let us take for example the parabola A P P', (Fig. 4), of which the equation is $y = x^2$, and suppose it to be composed of an infinite number of infinitely small straight lines. For the sake of perspicuity let us consider P P' as one of these sides; then representing A B by x, B B' by h, P B by y, and P' B' by y', we shall have

Fig. 4.

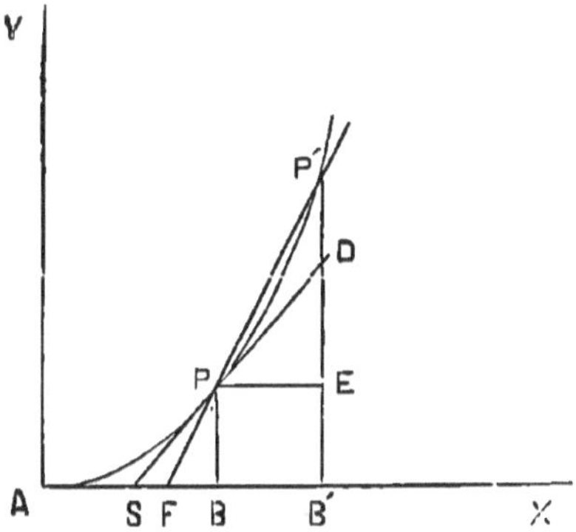

$$y = x^2 \text{ and } y' = (x + h)^2$$

whence we derive

$$\frac{y' - y}{h} = 2x + h$$

which represents the tangent of the angle P F B. But the true tangent of the angle made by the tangent line to a parabola like this is equal to $2x$; hence the prolongation of P P' is not tangent to the curve, and the P P' is not therefore a part of it. In considering the curve as made up of straight lines then, we commit an error, and this error we can correct only by reducing h to zero in the equation $\frac{y' - y}{h} = 2x + h$, and thus, bringing the two points P and P' together, we reduce the chord (or side of the curve) PP' to zero also, at the same time; so that while one of the postulates of de L'Hopital leads him into error, the other delivers him from it, and thus he is confirmed in both errors by the correctness of his results. "Two rays of light sometimes make darkness, but it is reserved for mathematics to exhibit the phenomenon of two rays of darkness producing light."

It is easy to show that the error arising from taking P′ P as a par: of the curve is exactly compensated by throwing out h. For if S D is the true tangent line, then instead of $\frac{P'E}{PE}$ we should have $\frac{DE}{PE}=\frac{PB}{SB}$ for the tangent of the angle made by the tangent line with the axis of abscissas. And since $SB=\frac{AB}{2}=\frac{x}{2}$ we have $\frac{PB}{SB}=\frac{y}{\frac{1}{2}x}=\frac{x^2}{\frac{1}{2}x}=\frac{DE}{PE}=2x$; but we found $\frac{P'E}{PE}=2x+h$, so that by throwing out h we have reduced $\frac{P'E}{PE}$ to $\frac{DE}{PE}$ and thus exactly compensated and canceled the error arising from assuming that the curve is composed of straight lines. We may therefore freely adopt the language of the celebrated Lagrange. He thus expresses himself on the subject in the last edition of his "Théorie des Fonctions Analytiques."

"In regarding a curve," says he, "as a polygon of an infinite number of sides, each infinitely small and of which the prolongation is the tangent to the curve, it is clear that we make an erroneous supposition; but this error finds itself corrected in the calculus by the omission which is made of infinitely small quantities."

Now this explanation of the reason for throwing out h was evidently unknown to Leibnitz himself; for when questioned on the subject he "presented" says Comte "an explanation entirely erroneous, saying that he treated infinitely small quantities as incomparables and that he neglected them in comparison with finite quantities 'like grains of sand in comparison with the sea'—a view which would have completely changed the nature of his analysis, by reducing it to a mere approximative calculus, which, under this point of view, would be radically vicious." Long before M. Comte, M. D'Alembert had said substantially the same thing, "M. Leibnitz" says he, "embarrassed by the objections which he felt would be made to infinitely small quantities, such as the Differential Calculus considered them, has preferred to reduce his infinitely small quantities to be only *incomparables*, which ruined the geometrical exactness of the calculus."

The first to discover the true reason for throwing out h, was Berkeley, Bishop of Cloyne.

"For as much" says he, "as it may perhaps seem an unaccountable paradox that mathematicians should deduce true propositions from false principles, be right in the conclusion, and yet wrong in the premises, I shall endeavor particularly to explain why this may come to pass, and *show how error may bring forth truth though it can not bring forth science.*"

He then gives a demonstration substantially similar to the one already given, and proving the same thing. But notwithstanding all this, together with the testimony of such men as have been named and many others, a large part of the teachers of the calculus practically adopt the method of Leibnitz; and when the puzzled pupil demands an explanation, he is told that the principle is true because the results are true. Not being able to deny this, the pupil contents himself with learning the use of the analytical machinery of the calculus, until, after a while, he forgets his doubts and acquiesces in the absurdities of the system.

A strong incentive to the continued use of the infinitesimal method is, that it is applicable to a great variety of cases, which other systems, heretofore developed, are unable to reach. The author of this work ventures to believe that he has presented a system of equal fecundity, based on sound principles, lying in the beaten track of mathematical thought

METHOD OF NEWTON.

A few years before the publication of Leibnitz' Differential Calculus, Sir Isaac Newton invented a process by which he obtained the same results, but the philosophy or fundamental idea of his system was wholly different.

In fact there are two systems, each of which goes by the name of *Newton's method.* The one generally in use is known as the "method of limits." By this method, in the equation $\frac{y' - y}{h} = 2x + h$, the quantity h is supposed to be reduced to absolute zero; and $2x$ is said to be the *limit* of the ratio of the increments of the variables, but not the ratio of any part of them, even infinitesimal. The results thus obtained are undoubtedly true, as will be seen in the case already considered. As the secant P F (Fig. 4) changes its position by the diminution of h or B B', the limit will be reached when h has become zero, P and P' have come together, P F has become the tangent line D S, and the tangent of the angle made by it with the axis of abscissas has become $2x$—the true result. Such is in substance the conception which Newton formed for the transcendental analysis: or, "more precisely, that which Maclaurin and D'Alembert have presented as the most rational basis of that analysis, in seeking to fix and arrange the ideas of Newton upon that subject." There is, however, one difficulty about it which has always puzzled the mathematical world. When h becomes equal to zero, the first member of the equation, $\frac{y' - y}{h} = 2x + h$, becomes $\frac{0}{0}$—a symbol of indetermination. The question that naturally arises is, how is the equation $\frac{0}{0} = 2x$ to be interpreted? It is one that has not hitherto been answered with that certainty of evidence which carries with it unhesitating conviction. The most plausible answers have been, in fact, only probable conjectures. Some persons assume that there is a real ratio between two zeros; others that the expression $\frac{0}{0}$ has no meaning: others, that while it has no real value, it is a "*symbol of a limit;*"—whatever that may mean—others, and these are the most numerous, without attempting to answer the question, or to give a reason for their proceeding, boldly substitute $\frac{dy}{dx}$ in the place of $\frac{0}{0}$ and assume that $2x$ is the ratio between the differential of the variable and that of its function. The following is an example of this kind, in a well known and popular work, recently revised and cured of its former inconsistencies:

* "It must not be supposed that the limiting value of a ratio is a mere *approximation;* for, while we speak of h approaching zero, we do this merely to aid our conception of the terms of the ratio; but in finding the value of the limit of the ratio, we actually take h equal to zero. The limiting value of the ratio thus found is, therefore, not approximately but absolutely exact."

"It is important," he says, "to have some convenient notation for expressing the limiting ratio of the two simultaneous increments. Now if dx be put for incr. x in the limiting value, then, in order to maintain a uniform notation, we must put dy for incr. y. We shall then have $\frac{dy}{dx} = 2x$. By $\frac{dy}{dx}$ we therefore understand the limiting value of $\frac{\text{incr. } y}{\text{incr. } x}$; or the limiting value of the ratio of the increment of y to the increment of x when the increment of x approaches zero."

But why substitute $\frac{dy}{dx}$ for $\frac{0}{0}$ since they are, in fact, identical in value? The reply is, that the author has use for dy and dx, where zero would not answer his purpose at all. For he says:

* Loomis' Elements of the Calculus, revised edition, pp 24 and 26.

"Not only does the symbol $\frac{dy}{dx}$ have an important meaning, but a meaning may be attached to the symbols dy and dx taken separately. These symbols may be regarded as representing the simultaneous *rates of increase* of a function and of the variable upon which it depends."

Now here is a most wonderful piece of logic! We may, of course, substitute for $\frac{0}{0}$ any two terms whose ratio is $2x$, whether we call them dy and dx, or a and b, or m and n; but by what legerdemain are dy and dx transformed into quantities, not merely having the constant ratio, $2x$, but endowed with the higher and more complex qualities by which they represent *rates*, which are relations between time and change; and this too by the mere stroke of a pen, without the trouble of a thought on the subject?

This assumption is characterized by Comte as an unwarranted mixture of the system of Leibnitz with that of limits; in reference to which practice, he remarks:

* "Several continental geometers, in adopting the method of Newton as the more logical basis of the transcendental analysis, have partially disguised this inferiority, by a serious inconsistency, which consists in applying to this method the notation invented by Leibnitz for the infinitesimal method, and which is really appropriate to it alone. In designating by $\frac{dy}{dx}$ that which logically ought, in the theory of limits, to be denoted by $L\frac{\Delta y}{\Delta x}$, and in extending to all other analytical conceptions this displacement of signs, they intended, undoubtedly, to combine the special advantages of the two methods; but, in reality, they have only succeeded in causing a vicious confusion between them, a familiarity with which hinders the formation of clear and exact ideas of either. It would certainly be singular, considering the usage in itself, that by the mere means of signs, it could be possible to effect a veritable combination between two theories so different as those under consideration."

* Philosophy of Mathematics, p. 238.

Dr. Bledsoe not only repudiates the infinitesimal method, but also the use of the signs dx and dy in the method of limits. He says, referring to the expression $\frac{0}{0}$:

"It is said, if we retain this symbol, our operations may be embarrassed or spoiled by the necessity of multiplying, in certain cases, both members of an equation by zero. But the answer is easy. The first differential coefficient, if rendered accurate, always comes out in the form of $\frac{0}{0}$; but it need not retain that form at all. Whether we use $\frac{0}{0}$ or $\frac{dy'}{dx'}$ in writing the differential equation of a tangent line to the point x', y', we shall have to eliminate $\frac{0}{0}$ in the one case, and $\frac{dy'}{dx'}$ in the other, in order to make any practical application of the formula. Now $\frac{0}{0}$ is just as easily eliminated by the substitution of its value in any particular case as is $\frac{dy'}{dx'}$, and besides its value may be found and its form eliminated by substitution without any false reasoning or logical blunder, which is more than can be said for the form $\frac{dy'}{dx'}$."

This view would greatly restrict the operations of the calculus, confining them to the solution of the most simple problems in it and leaving those of a more intricate kind, as well as those of the Integral Calculus and most of those in mechanics, without the means of solution altogether. It was this inherent weakness of the method of limits that induced the earlier mathematicians to assume, without giving reasons or explanations, the symbols used in the infinitesimal method, endeavoring, at the same time, to conceal their essential identity by attaching a different meaning to the term "differential." This, of course, enlarged the scope of their system, but still left it unable to cope with the more intricate problems of mechanics. It is now, however, used by very many teachers, but hitherto no logical demonstration has been given to justify the assumptions which contain the essence of the system. It may here be remarked that neither the system under consideration, nor the pure method of limits, was ever adopted by Newton as his method of constructing the calculus. His only connection with the method of limits was, the demonstration of certain principles applicable to that method, but which he used for a very different purpose. Nevertheless, this compound system, which has been so unceremoniously adopted, produces true results; and, notwithstanding the hitherto lack of rigid, mathematical proof of its soundness, it has at the bottom a profound truth, which, being found, will be a lamp to our feet to guide us out of the region of doubt and perplexity enveloped in the mists of metaphysical speculation, into the pure light of reason and logical certainty. This principle is that which lies at the foundation of Newton's true system, which he invented in the year 1666, and which was afterwards published under the title of *fluxions*. This *method* we will examine in due time; but meanwhile we will proceed to show what this principle is, and how it applies to the elucidation of the method under consideration.

PRIME AND ULTIMATE RATIOS.

One of the conditions, under which the variable quantities used in the calculus are supposed to exist, is, that they are not only capable of assuming different values, but are *actually in a state of change.* The intensity of this state at any instant is measured by the *rate* at which the variable may be changing at that instant; and the symbol which measures this rate, at any moment, is the change that would take place in a unit of time, if the rate were to continue uniform, from that moment, during that unit; and the ratio of two rates is the ratio of the simultaneous changes that *would* take place, if the rates were to continue uniform for *any length of time.* It is this state or rate, as represented by the symbol just mentioned, and not any value, nor actual change of value, in the quantity, which we represent by its differential.

Another of the peculiar conditions attached to variables is, that they are susceptible of two opposite relations to the point from which their value is estimated — that is, from the zero point: they may be essentially negative or positive, independent of the algebraic sign which precedes them; and the same quantity may pass, continuously, from a positive to a negative state, or *vice versa*, through the zero point which divides them, without losing, at that point, all its characteristics as a quantity. This point of separation is generally arbitrary: as the zero point in latitude, the axes of coördinates in geometry, etc., etc. Now the rate, at which any such quantity may be changing, is not at all affected when its value becomes zero, as it passes from an essentially positive to an essentially negative state, or *vice versa;* but will be governed by the same law at that point as that which controls it elsewhere. Thus the cosine of an angle of 120° is essentially negative, independent of the sign which precedes it; and in its change from the point where the angle begins, it will diminish from positive radius to zero, and then become negative, while the *rate* of change, if determined by a uniform increment of the angle, will be greatest at the point where the cosine itself is nothing.

Rate is a complex idea of which the elements are time and change; but when one uniform rate is compared with another, without regard to absolute values, the idea of time may be left out of the question, and the comparison may be made by that of their simultaneous changes; so that the ratio of the simultaneous increments of any two variables which increase at constant rates, does in fact perform a *double office;* representing not only the ratio of their changes, but also that of *their rates of change*, which will of course be constant.

To illustrate this let us take a very simple case:

Let $y = ax$. Then by giving to x an increment, h, and denoting by y' the increased value of y, we have

$$y' = ax + ah,$$

and subtracting and dividing by h, we have

$$\frac{y' - y}{h} = a.$$

If now we suppose x to be increasing at a uniform rate, then h and $y' - y$ will be the uniform simultaneous increments of y and x, and will express their *relative* rates of increase; and these rates being constant, their ratio will be constant, and equal to the constant ratio (a) of the increments, irrespective of any particular value of h. Now, if we suppose h to decrease, the value of the ratio, expressed by the second member of the equation, will not change, even when h becomes zero and the fraction assumes the form $\frac{0}{0}$. How then are we to interpret the expression

$$\frac{0}{0} = a?$$

Evidently, since a has been the ratio, not only of the increments, but also of their rates of increase, and since we have seen that the rates do not necessarily vanish with the values of the quantities, the constant ratio, a, must be that of the *rates of change* in the increments, as they are passing through the zero point of their value. Hence the expression $\frac{0}{0}$, which is commonly supposed to be meaningless, and which, when isolated, is so in fact, has, when it arises as the result of an analytical process, a very definite and distinct meaning, which is to be derived from the conditions which produce it. Hence we infer that $\frac{0}{0}$ derived from the equation $\frac{y' - y}{h} = a$ means that the quantities represented by the terms of the fraction *are then passing through the point where they would change from positive to negative*, and that while their values have disappeared, their rates of change are not at all affected; and since a rate of change implies a changing quantity, we can not consider these zeros, *thus derived*, as absolutely nothing in all respects; but while they have no *actual value*, as quantities, they have a *character*, and must be treated as quantities in reference to the characteristics which remain to them when their *value* has departed — and these characteristics are, their *state of change*, and *rate of change*; and a must be considered as the ratio of their rates of change at that point, as well as when it was also the ratio of their values.

Let us now resume the equation $\frac{y' - y}{h} = 2x + h$. In this case the terms of the fraction do not change uniformly, and therefore do not express their respective rates of change; so that this equation can not be interpreted as in the last case. The ratio $2x + h$ must, however, be a function of the rates, since it depends upon them for its value; let us examine the conditions of that dependence.

We have seen that, when we reduce h to zero and obtain $\frac{0}{0} = 2x$, which is the ratio with which the increments vanish, the fraction is not necessarily an unmeaning form of expression; and hence we have a right to look for some definite and rational interpretation of it — in other words to seek, with the hope of an intelligent answer, of what quantities is $2x$ the ratio?

In discussing this subject our first proposition is, *the ratio with which the increments vanish is, also, that with which they begin to be.*

This proposition is so nearly self-evident that Sir Isaac

Newton did not hesitate to assume it without a demonstration. It may, however, be made more apparent by a little variation of the process we have been using. In the equation $y = x^2$, let x be *decreased* by the quantity h, and then let it increase from $x - h$ to $x + h$ (h changing its sign at the zero point) and we shall have these three forms of the equation, viz. $y'' = (x - h)^2$, $y = x^2$ and $y' = (x + h)^2$ from which we derive.

$$\frac{y - y''}{h} = 2x - h, \qquad (1.)$$

$$\frac{0}{0} = 2x, \qquad (2.)$$

$$\frac{y' - y}{h} = 2x + h. \qquad (3.)$$

As h diminishes in equation (1), the ratio of the decrements will increase, until h vanishes, and y'' being now equal to y, we have equation (2). Here h changes its sign and begins to increase, completing the continuous increment of x from $x - h$ to $x + h$ and producing equation (3).

The equation $\frac{0}{0} = 2x$ is evidently derived from the ratio of the decrements, just as it was from that of the increments, and is in fact the point where the former end and the latter begin. Hence, in equation (2), we may consider $2x$ as the ratio with which the increments begin.

Our second proposition is, that the ratio $\frac{0}{0} = 2x$ *is not* the ratio of any part of the increments, h and $y' - y$. It is just here that we meet and deny the infinitesimal theory, which is that $2x$ is the ratio of the last values of the increments as they vanish—that is to say it is the ratio of the infinitely small portions of the increments. If authority were of any avail in such a case as this, we have that of Sir Isaac

Newton most directly and emphatically upon this very point.* But no authority is needed to establish the truth, that $2x + h$ is not equal to $2x$ as long as h is *anything*—we *must* think so if we think at all. In fact the infinitesimal theory would probably never have been thought of, had it not seemed to be necessary to account for the ratio $2x$, by supposing that there *must* be some sort of quantities to be compared, and infinitesimals were invented to supply the demand.

The theory can no longer be admitted on this ground, for we have seen that, in a similar case, the zeros are not absolute nonentities. They have indeed lost one of the characteristics of quantity, viz., that of *value*, but have retained those of *change* and *rate* which will readily furnish quantities to be compared, and thus leave the infinitesimal theory, with all its inconsistencies and absurdities, without the shadow of a reason for its existence.

We now come to the question — of what is $2x$ the ratio? The increments must, of course, begin with the same rates of increase as those with which the variables themselves were increasing at that moment, and before the increments had acquired any magnitude; and these rates will be represented by the increments that *would take place* in any unit of time should the rates continue to be uniform. But if the rates were to continue uniform, their ratio as well as that of the increments would be constant, and would be that with which the increments began, and with which the variables were increasing at the same moment. It is true that in the case under consideration the increments are not formed at a uniform rate, and their ratio, therefore, is not constant, and is not that of their rates of change; but it is also true that $2x$ is *what would be* their constant ratio if they did increase uniformly from the beginning; and these suppositive uniform increments, as we have seen, are *the very symbols*

*Principia, Book I, Lemma IX. Scholium.

which measure the rates with which the increments begin; or what is the same thing, the rates with which the variables were increasing at the same moment, whatever changes the rates may undergo afterwards. Hence $2x$ — the ratio with which the increments begin — is also the ratio of the suppositive uniform increments or symbols which measure the rates of change in x and y (or x^2) at the moment the increments begin, and is therefore THE RATIO OF THE RATES THEMSELVES.

The abstract proposition may be thus demonstrated.

PRIME AND ULTIMATE RATIOS.

1. THE RATIO OF *UNIFORM* INCREMENTS IS THAT WITH WHICH THEY VANISH AND, ALSO, BEGIN; AND IS ALSO THE RATIO OF THEIR RATES OF INCREASE.

2. THE RATIO WITH WHICH *VARIABLE* INCREMENTS VANISH IS THAT WITH WHICH THEY BEGIN; AND IS ALSO THE RATIO OF WHAT *WOULD BE* THE INCREMENTS IF THEY WERE MADE AT A UNIFORM RATE.

3. THESE SUPPOSITIVE, SIMULTANEOUS, UNIFORM INCREMENTS ARE SYMBOLS, WHICH REPRESENT THE INITIAL RATES OF INCREASE OF THE REAL INCREMENTS.

4. HENCE THE ULTIMATE RATIO (WHICH IS ALSO THE NASCENT RATIO) OF THE INCREMENTS IS THE RATIO OF THE SYMBOLS WHICH REPRESENT THE RATES WITH WHICH THE INCREMENTS BEGIN: AND THESE ARE THE RATES AT WHICH THE VARIABLES THEMSELVES ARE CHANGING AT THAT MOMENT: THAT IS TO SAY — THEY ARE DIFFERENTIALS OF THOSE VARIABLES; AND THE ULTIMATE RATIO OF THEIR VANISHING INCREMENTS IS THEIR TRUE DIFFERENTIAL COEFFICIENT.

If now we represent these indeterminate symbols of the rates, by the signs dx and dy (calling them differentials of x

and of y) we may substitute these terms in the fraction and write $\frac{dy}{dx}=2x$ instead of $\frac{0}{0}=2x$, an equation which is always true, expressing, not the actual values of the terms of the fraction, but the *relation* existing between the differential, or rate of change in x, and the corresponding rate of change in y or x^2.

We have therefore a right, notwithstanding the opinions of Comte and Dr. Bledsoe, to substitute $\frac{dy}{dx}$ for $\frac{0}{0}$, in what is called the method of limits, provided we understand by them, not infinitely small quantities nor zeros, but *symbols*, which represent the simultaneous rates of change in y and x, and whose *absolute* values, like those of other variables, are indeterminate until we introduce the conditions necessary to fix them.

We may take another view of the subject which will lead to the same result.

When any variable is in a state of change at a variable rate, we may always consider the actual change occurring during any unit of time, as arising from two distinct causes; namely, partly from the rate *existing* at the beginning of the time, and partly from the *change* in the rate taking place during that time. So if a body is falling to the earth, the space passed over during one second, is composed of two distinct parts; one being that which it would have passed over at a uniform rate by virtue of its *initial* velocity at the beginning of the time, and which would be the measure of that velocity; and the other, that passed over by virtue of the *increased* velocity produced by the action of the force of gravity. Hence if we take the ratio of the simultaneous increments of two variables increasing at variable rates, this ratio will be affected by the same causes that control the values of the increments themselves; namely, partly by the rates existing when the increments began, and partly by the subsequent changes in those rates; and any change in the ratio of the increments will be due wholly to the change in the rates which may take place while the increments are forming. For if the increments were the result of only the rates existing at their beginning, they would be made at a uniform rate, and their ratio would be constant, and, of course, would be the ratio of the rates with which they began and continued; but if certain causes should change the rates of these increments, it would ordinarily be manifested by a change in their ratio.

Now, as we diminish the increments, we diminish the changes in their rates, and finally when they disappear, the changes in their rates also disappear, — and of course the changes in their ratio vanish at the same time, leaving it just as if there had been no changes in the rates — that is, it will be the ratio of the rates with which the variables were increasing when the increments began, and before any changes in the rates occurred.

Thus in the equation $\frac{y' - y}{h} = 2x + h$, the quantity h in the second member is the change in the ratio produced by change in the rate of y; and if we reduce the increments to zero the changes in the ratio will disappear and the ratio $2x$ will be the same as if x and y were increasing at a uniform rate while the increments were forming, and is therefore the ratio of their rates of increase with which they begin.

Thus we see, the truth of the method of ultimate ratios (commonly called the method of limits) is shown to rest wholly on the doctrine of rates represented by concrete symbols, or suppositive uniform increments; and these are what are known as differentials. This being so it follows that the differentials of the calculus are not infinitesimals, but *rates* of *change* symbolically expressed, and that the calculus itself has nothing to do with infinities nor metaphysics of any kind, but is simply, THE SCIENCE OF RATES.

This method of final ratios although logically true is yet very inferior to the *direct* method of rates, for

First, It is indirect. That is never so satisfactory as a direct philosophical method; and when such a method can be found (as it has been in this case), if other things are equal, it is always to be preferred.

Second, But other things are not equal, but are strongly in favor of the direct method. The method of limits, requiring an addition and then a diminution of increments, imposes on the mind the conception of a process utterly useless, besides producing a false conception of the true nature of a variable quantity and of the philosophical relation between it and its differential; for the process completely ignores the essential characteristic of a true variable.

Third. This method is powerless to solve many problems which have hitherto been solved only by the use of infinitesimals, whereas the direct method is fully equal in scope to the infinitesimal system, its analytical processes being generally the same, while the idea suggested by those processes, instead of being absurd and false, is true and philosophical. Thus, instead of saying, the infinitesimal increment of a surface is equal to the infinitesimal increment of x multiplied by the ordinate y, we may say the rate of change of the surface is equal to the rate of change of x multiplied by y. In either case we have $ds = ydx$. The first idea is incomprehensible, while the latter is plain to the dullest comprehension.

Fourth. The demonstration required to prove the truth of the method of final ratios is by no means an easy one to comprehend; and the method itself fails to keep prominently before the mind the idea of the differential as represented by the concrete symbol, and leaves the student to feel that he is working in a region where all is shadowy and unsubstantial. It is like traveling over a rough road in the dark, guided, it is true, by an unerring hand, compared with traveling over the same road in the broad light of day.

The boast of the infinitesimal system is, that it "presents incontestably, in all its applications, a very marked superiority, by leading in a much more rapid manner, and with much less mental effort, to the formation of equations between auxiliary magnitudes."*

This arises from the fact that the method is *direct.* Instead of seeking, by the method of final ratios, the differential coefficient, and finding from that the differential itself, the infinitesimal method seizes the differential at once as an infinitesimal increment; making the differentials of all curves to be infinitesimal straight lines; the velocity of variable motion to be uniform during an infinitesimal portion of time, while the body moves over an infinitesimal portion of space.

Hence, since in uniform motion we have $s = vt$, the equation will hold good for variable motion when the time and space are infinitesimal, and we have $ds = vdt$, or $v = \frac{ds}{dt}$ which is thus obtained directly with perfect ease.

Now these advantages of the infinitesimal method are incontestible; and were the principles of that method true, there would be nothing left to be desired. But its principles are false. No curve is made up of straight lines — no velocity is uniform in a variable motion; and hence if a method can be found which is equally applicable to every variety of investigation, equally direct, and requiring no more mental power for its conception, and yet rigidly sound in its principles and logical in its application, there should be no hesitation in adopting it.

* Comte Phil. of Math., p. 113.

Now the method of rates, expressed by uniform symbolical increments, is such a method. The analytical machinery is precisely the same as that of the infinitesimal method, and it differs from that system only in the conception formed of the nature of the differential. The latter method conceives it to be an *actual, infinitesimal, uniform* increment of the variable, while the former conceives it to be a *suppositive, imaginary, symbolic, uniform* increment, which performs exactly the same duty. Thus instead of velocity being represented by an infinitely small space passed over in an infinitely small portion of time, it is represented by a suppositive space that *would be* passed in a suppositive unit of time, if the velocity were to continue uniform during that unit; which is in fact the method by which we always practically measure velocity: as when we say "a body is falling to the earth with a velocity of 50 feet in one second." In this case the suppositive space and suppositive time are the differentials, and we have, as in the other, $ds = vdt$ or $v = \frac{ds}{dt}$.

THE TRUE METHOD OF NEWTON.

I have said that the method of arriving at the differential coefficient by means of the ultimate ratios of the increments, or, in other words, the method of limits, has generally been ascribed to Sir Isaac Newton: but this is evidently an error. The *theory* on which that method is founded is certainly his, and it is but just that he should be held responsible for the results that legitimately flow from it. But it is not the theory on which he formed *his* method of fluxions. *That* is contained in the second lemma of the second book of his Principia. In a scholium to that lemma he says: "In a letter of mine to Mr. J. Collins, dated Dec. 10, 1672, having described a method of tangents — which at that time was made public, I subjoined these words. *This is one particular or rather corollary, of a general method, which extends itself, without any troublesome calculation, not only to the drawing of tangents to any curved lines, whether geometrical or mechanical, or anyhow resolving other abstruse kinds of problems about the crookedness [curvature] areas, lengths, centers of gravity of curves, etc., nor is it limited to equations which are free from surd quantities. This method I have interwoven with that other of working equations, by reducing them to infinite series.* So far that letter. And these last words relate to a treatise I composed on that subject in the year 1671. The foundation of that general method is contained in the preceding lemma."

Here it is distinctly stated by Newton himself that he had invented a *general method* which was applicable not only to the drawing of tangents, but to all the higher and more delicate problems which appear in the Differential Calculus, and that this general method has *the lemma in question for its* FOUNDATION.

We have then but to examine this lemma to ascertain the real basis on which the "method of Newton" was constructed. For this purpose we give the lemma in the author's own words.

LEMMA II.

"*The moment of any genitum is equal to the moments of each of the generating sides drawn into the indices of the powers of those sides, and into their coefficients continually.*

"I call any quantity a genitum which is not made by the addition or subduction of divers parts, but is generated or produced in arithmetic by the multiplication, division or extraction of the root of any terms whatsoever; in geometry by the invention of contents and sides, or the extremes and means of proportionals. Quantities of this kind are products, quotients, roots, rectangles, squares, cubes, square and cubic sides and the like.

"These quantities I here consider as variable and indetermined, and increasing or decreasing as it were by a perpetual motion or flux; and I understand their momentaneous increments or decrements by the name of moments: so that the increments may be esteemed as additive or affirmative moments, and the decrements as subducted or negative ones. But take care not to look upon finite particles as such. Finite particles are not moments, but the very quantities generated by the moments. We are to conceive them as the just nascent principles of finite magnitudes. Nor do we in this lemma regard the magnitudes of the moments, but their first proportion as nascent. It will be the same thing, if, instead of moments, we use either the velocities of the increments and decrements (which may be called the motions, mutations, and fluxions of quantities) or any finite quantities proportional to those velocities. The coefficient of any generating side is the quantity which arises by applying the genitum to that side.

"Wherefore the sense of the lemma is, that if the moments of any quantities A, B, C, etc., increasing or decreasing by a perpetual flux or the velocities of the mutations which are proportional to them, be called a, b, c, etc., the moment or mutation of the generated rectangle AB will be $aB + bA$; the moment of the generated content ABC will be $aBC + bAC + cAB$; and the moments of the generated powers A^2, A^3, A^4, $A^{\frac{1}{2}}$, $A^{\frac{3}{2}}$, $A^{\frac{1}{3}}$, $A^{\frac{2}{3}}$, A^{-1}, A^{-2}, $A^{-\frac{1}{2}}$. will be $2aA$, $3aA^2$, $4aA^3$ $\frac{1}{2}aA^{-\frac{1}{2}}$, $\frac{3}{2}aA^{\frac{1}{2}}$, $\frac{1}{3}aA^{-\frac{2}{3}}$, $\frac{2}{3}aA^{-\frac{1}{3}}$, $-aA^{-2}$, $-2aA^{-3}$, $-\frac{1}{2}aA^{-\frac{3}{2}}$, respectively: and in general that the moment of any power $A^{\frac{n}{m}}$ will be $\frac{n}{m}aA^{\frac{n-m}{m}}$. Also that the moment of the generated quantity A^2B will be $2aAB + bA^2$; the moment of the generated quantity $A^3B^4C^2$ will be $3aA^2B^4C^2 + 4bA^3B^3C^2 + 2cA^3B^4C$; and the moment of the generated quantity $\frac{A^3}{B^2}$ or A^3B^{-2}, will be $3aA^2B^{-2} - 2bA^3B^{-3}$, and so on. The lemma is thus demonstrated.

"*Case* 1. Any rectangle, as AB, augmented by a perpetual flux when as yet there wanted of both sides A and B, half the moments $\frac{1}{2}a$ and $\frac{1}{2}b$, was $A - \frac{1}{2}a$ into $B - \frac{1}{2}b$, or $AB - \frac{1}{2}aB - \frac{1}{2}bA + \frac{1}{4}ab$; but as soon as the sides A and B are augmented by the other half moments, the rectangle becomes $A + \frac{1}{2}a$ into $B + \frac{1}{2}b$, or $AB + \frac{1}{2}aB + \frac{1}{2}bA + \frac{1}{4}ab$. From this rectangle subduct the former rectangle, and there remains the excess $aB + bA$. Therefore with the whole increments a and b of the sides, the increment $aB + bA$ of the rectangle is generated. Q. E. D."

"*Case* 2. Suppose AB always equal to G, and then the moment of the content ABC or GC (by case 1) will be $gC + cG$, that is (putting AB and $aB+bA$ for G and g) $aBC+bAC+cAB$. And the reasoning is the same for contents under ever so many sides. Q. E. D."

It is unnecessary to quote the demonstrations of the other cases, as they all flow naturally and logically from these which form the key to the whole system.

We must concede that this demonstration is not as clear and complete as could be desired. Let us, however, endeavor to extract from it the real, though perhaps somewhat vague conception of the subject which occupied the mind of Newton. It is to be remarked, however, that the doctrine of *limits* is nowhere hinted at, but the results are direct, positive and substantial.

The first question suggested by the lemma is, what is really meant by the term "moment." It might at first seem that the "moments" of Newton were in fact the same thing as the differentials of Leibnitz, for he speaks of them as something (though not finite quantities) to be added or subtracted. But a very little examination of the lemma will dispel the notion. Their magnitudes are not to be regarded. But the magnitudes of the differentials of Leibnitz *are* to be regarded as infinitely small. Again, "finite particles" are not "moments," but the "very quantities generated by the moments." Now the differentials of Leibnitz never generate any thing; they are the infinitesimal remains of increments that have been added and then taken away. Again, moments are the "nascent principles of finite magnitudes." But the "principles" which generate "finite magnitudes" or increments can be nothing else than the *laws* which control the changes in the "genitum;" that is, THE RATE OF CHANGE. This interpretation is confirmed by the further statement that we may use instead of them "the velocities" or any finite quantities proportional thereto. Hence we infer that a, b, c, which are called moments, are intended as *symbols* to represent the rates of change, being finite quantities proportional to those rates, and as the quantities A, B, C, etc., are increasing or decreasing by a "perpetual flux," that is by a uniform rate of change, the actual increments or decrements *a*, *b*, *c* will represent those rates. So that the difference between $A-\frac{1}{2}a$ and $A+\frac{1}{2}a$ (equal to *a*) represents the rate of increase of A, and the difference between $B-\frac{1}{2}b$ and $B+\frac{1}{2}b$ (equal to *b*) accruing during the same time represents the corresponding rate of increase of B; and the ratio of *a* to *b* represents the ratio of those rates whatever may be their magnitude as symbols. But while these symbols or suppositive increments (being produced at a uniform rate) represent the respective rates of increase of A and B, we are told that the corresponding increment of their product ($aB+bA$) represents the "moment" or rate of increase of their product. Now as the product does not increase at a uniform rate, it becomes a question why *this increment* should represent the *rate* of increase of AB. This is probably one of those cases in which the intuitive perceptions of Newton seized the true result without stopping to elaborate the intermediate steps.

The solution of this question will be found in this work.

We have then, for the true method of Newton, the rates of change of the variables, instead of the infinitely small differences between their increments. It is an interesting question, why this method has not been more generally used by the teachers of the Calculus.

One reason is probably, that although it was discovered by Newton as early as 1666, it was not, for many years, put into systematic form for public use; not, in fact, for any purpose, until 1671, and then remained dormant, being only referred to in a lemma in his Principia published in 1687. In that he merely pointed out the principles, and did not himself give any thing else to the world until, in 1704 he published a tract explaining the principles of his method and applying it to quadratures. Finally, in 1711, Prof. Barrow published a treatise which Newton had put into his hands, containing a general method of calculating the quadratures of curves, and also resolving equations. Thus the real method of Newton came gradually before the world, in such a shape as not to attract general attention, nor to be practically used for making discoveries in mathematics. His attention was, no doubt, so intensely occupied by his multitudinous labors in other departments of science, that he did not himself, realize the importance of his discovery.

Meanwhile Leibnitz had, in 1684, published his Differential Calculus in which he gave rules for applying his method, so that it was ready for immediate public use. It attracted attention at once, not merely by its utility in the solution of difficult problems, but also by the new and extraordinary conceptions it contained as to the nature of quantities, which conceptions could not be denied since the results founded on them were found to be true. All these circumstances combined to give to the method of Leibnitz a precedence which it has continued to hold from the beginning.

The invention of what is known as, "*the method of limits*" had, also, an influence in keeping the true method of Newton in the back ground. This method is founded on the following lemma in the first Book of his Principia.

"*Quantities and ratios of quantities, which, in any finite time converge continually to equality, and before the end of that time approach nearer the one to the other than by any given difference, become ultimately equal.*"

From this circumstance, when it was formulated and brought into use, it was termed "Newton's method," and thus obtained the prestige of his great name. It was at once seized by those whose reason was shocked by the absurdities of the infinitesimal system, since, although it was feeble, it had no logical inconsistencies. To strengthen it, some geometers adopted the symbols of Leibnitz, and, in this way, endeavored "to combine the advantages of the two methods"; and, when the results were found to be true, were confirmed in the use of this compound method, without insisting on a logical demonstration of the principle they assumed. It put to rest the feeling of inconsistency, arising from the use of infinitesimals, and they were content.

Another reason for the neglect of Newton's true method, may be found in the symbols he adopted, which were not well adapted to analytical operations.

"* The variables are denoted by the final letters of the alphabet; as x, y, z etc. and their fluxions are indicated by the same letters with a dot over them. Thus $\dot{x}, \dot{y}$ and $\dot{z}$ are the symbols of the fluxions of x, y and z. If the fluxions are variable, they may be regarded as fluents, whose fluxions may be taken, and these are denoted by the same letters with two or more dots over them, according to the order of the fluxion. Thus $\ddot{y}, \dddot{y}, \ddddot{y}$ etc. denote fluxions of y of the second, third, fourth etc. orders. If the fluent is a radical, as $\sqrt{x-y}$, its fluxion is denoted by placing the radical in a parenthesis, and writing a dot over it to the right, as $(\sqrt{x-y})^{\cdot}$. Also, the fluent of a fraction is written in a similar manner, thus, the fluxion of $\frac{x}{y}$ is written $\left(\frac{x}{y}\right)^{\cdot}$

"Sometimes the fluxion is indicated by the letter F, and the fluent by f; thus $F\,(\sqrt{x-y})$, is the same as $(\sqrt{x-y})^{\cdot}$.

"Also the expression of $f\dot{x}\sqrt{a-bx^2}$ and $f\left(\frac{b\dot{x}}{a+x^2}\right)$ denote the fluents of $\dot{x}\sqrt{a+bx^2}$ and $\frac{b\dot{x}}{a+x^2}$ respectively.

"This notation is exceedingly cumbrous, particularly in the higher branches of analysis, and for this reason, principally, the method of fluxions has gone into disuse."

Probably, however, the most effective reason why the true method of Newton has been suffered to die out, was,

* Davies & Peck — Mathematical Dictionary.

that its real essence was not thoroughly comprehended. Indeed it would seem that, in Newton's own mind, it was not perfectly clear and exactly defined. This is apparent in the extract we have given from the lemma in which he lays the foundation of his method. We have endeavored to extract from that lemma what was his real, although somewhat vague idea; and if the effort has been successful, it seems to be quite certain that most of his commentators have failed to do so. In order to have the subject fairly before our minds, let us re-state that principle, viz. — the function which Leibnitz terms "*differential*," and which Newton designates as a "*fluxion*," is the concrete *symbol* which represents the rate of change in the variable, and which consists of the change of that *would* take place at that rate, continued uniformly for one unit of time. It is not contended that this definite and plain statement was made by Newton; but it is believed that this was the true idea that was at the bottom of his method of fluxions.

The following extracts from some of Newton's commentators will show how widely their ideas differ from the one we have just stated. The following is from Davies & Peck's Mathematical Dictionary.

"The idea of fluxions and fluents was first presented by Newton and was based upon the idea of motion. According to his view a plain curve or line may be conceived as generated by a point moving uniformly in the direction of some fixed line, and having at the same time a lateral motion with respect to this line, which is governed by some law dependent on the nature of the curve generally. The part of the curve generated at any instant of time, is called the fluent, and that infinitely small element, generated during the next infinitely small and constant period of time, is called its fluxion."

We also find in Comte's Philosophy of Mathematics the following remarks on the fluxional method of Newton.

"It is easy," he says, "to understand the general and necessary identity of this method with that of limits, complicated with the foreign idea of motion. In fact, resuming the case of the curve, if we suppose, as we evidently always may, that the motion of the describing point is uniform in a certain direction, that of the abscissa, for example, then the fluxion of the abscissa will be constant, like the element of time; for all other quantities generated, the motion can not be conceived to be uniform except for an infinitely small time. Now the velocity being, in general, according to its mechanical conception, the ratio of each space to the time employed in traversing it, and this time being proportional to the increment of the abscissa, it follows that the fluxions of the ordinate, of the arc, of the area, etc., are really nothing else (rejecting the intermediate consideration of time) than the final ratios of the increments of these different quantities to the increment of the abscissa. This method of fluxions and fluents is, then, in reality, only a manner of representing by a comparison, borrowed from mechanics, the method of prime and ultimate ratios, which alone can be reduced to a calculus."

The following extract from Prof. Loomis' History of the Calculus which forms the introduction to the last revised edition of his work on that subject, would seem to show that he viewed the subject in the same light as Davies and Comte. He says:

"Indeed the fluxions and fluents of Newton correspond essentially to the differentials and integrals of Leibnitz; so that the two methods differ only in the notation, and in the peculiar modes of viewing the subject, or what is commonly called the *metaphysics* of the Calculus."

In reference to these extracts we remark — *First.* That the idea of the generation of a geometrical magnitude by the flowing of an element, which has been ascribed to Newton as the inventor, and which it is said suggested his method of fluxions, is by no means peculiar to the calculus. It is freely used in Descriptive and Analytical Geometry, and is moreover, in many cases, the best, and sometimes the only clear and concise method, of defining magnitudes; as in the case of surfaces and solids of revolution.

Second. That the differential, or fluxion, of a curve, is "that infinitely small element generated during the next infinitely small and constant period of time" is simply another mode of stating the infinitesimal theory, which is certainly not to be found in Newton's explanation of his own method. On the contrary it is clear from a careful examination of it, that there is nothing in it involving the idea of infinitesimals nor of limits. Such a statement then could have proceeded only from an entire misapprehension of the subject. M. Comte seems to think that he is forced to the same conclusion as Dr. Davies, from the fact, that since velocity can be measured only by comparing the space passed over at a uniform rate with the time occupied in traversing it, the space and time must be infinitely small in order to find, at any moment, the value of a variable velocity; and hence it can be found, only by the method of limits. If his premises were true, his conclusion would be reasonable. But it is *not* true that variable velocity is so measured; and it is surprising that a philosopher should forget that measure of velocity which is used by the whole world, every day, in common conversation. Namely, *the space that would be uniformly passed over in any unit of time* —at the same rate; as when we say a body is falling to the earth with a velocity of fifty feet in one second of time. Such a statement could have been made, only for the purpose of sustaining a pre-conceived theory founded on an entire misconception of Newton's method. These samples of this misconception are taken from modern writers, but they have only followed the notions that have been current in the mathematical world from the beginning, and prove that up to their time the principle of Newton's method had not been apprehended.

I have thus endeavored to show that the true method of Newton has been misunderstood, but if, in fact, the *writer* has mistaken the method of Newton, and his commentators have comprehended it aright, then the following work comprises A NEW METHOD, and can stand on its own merits without the support of Newton's great name. The reader is at liberty to form his own opinion as to the paternity of the "*direct method of rates;*" if he is unwilling to ascribe it to the great genius of the seventeenth century, the writer will cheerfully stand "in loco parentis."

PART I.

Differential Calculus.

DIFFERENTIAL CALCULUS.

SECTION I.

DEFINITIONS AND FIRST PRINCIPLES.

VARIABLES.

(1) Two classes of quantities are considered in the differential calculus, namely, *variables* and *constants*..

Variables are quantities that are in a state of change; that is, their values are in an increasing or decreasing condition; such, for example, as the quantity of water in a vessel which is being filled or emptied by a continuous stream; or as the force of attraction which increases or diminishes as the attracting bodies approach or recede from each other; or as the space between these same bodies while they are moving. They are, in the differential calculus, not merely quantities *subject* to change, or to which different values may be assigned; but quantities in which *the change is supposed to be actually occurring* at the moment when they become the subject of the analysis. It is their actual *condition* and not their attributes or qualities that are referred to in this definition. Take for example the space passed over by a falling body. That space is a variable, not because it may or does have different values, but because its value is *constantly changing*, or is in a *state of change*. It is this *state*, and not any actual change, that is the peculiar subject of the transcendental analysis.

RATE OF VARIATION.

(2) Rate of variation is the *relation* between the *change* of a variable and the *time* occupied by the change. Being a relation and not a simple quantity, it can only be represented by a *symbol*, which is *a uniform change in a given unit of time*. If the rate is constant, then the actual change is the true symbol, but if it is variable, then the change must be a *suppositive* one — that is, *one that would take place in the same unit of time if it were to continue uniform*. Thus in the case of a falling body, it is said the velocity at a certain moment is so many feet in one second. It is not meant that the body actually falls through that distance in one second, nor any distance whatever at that rate, but that it *would fall* so far if the velocity existing at that moment were to continue uniform for one second. The velocity belongs to that one moment, and that one position only. At the very next point above and below this position the velocity is different, and hence no actual movement, however small in respect to space and time, can possibly represent it. This will be seen at once if we consider what velocity is. It is not of itself a quantity, but a *relation*, which refers, not to the place, but to the *condition* of the body in respect to the motion — that is, to the *degree* or *intensity* of the state of motion in which the body is.

So it is with all variables. The rate of change refers to the intensity with which the change is going on, and if it is not uniform it can not possibly represent the rate, for it lacks the essential element of the required symbol. The latter must therefore be obtained from the *law* which governs the change and not from the change itself. Hence instead of giving to a function an actual increment for the sake of obtaining its rate of increase at any moment, we examine the *law* which governs the change; and the expression of this law is *the change that would take place in a unit of time if the rate were to continue uniform*; AND THIS IS THE MEASURE OF THE RATE.

NOTE.—It must be remarked that the ideas of time, motion and velocity, attached to the ordinary meaning of these words, have no place in the abstract science of the differential calculus. The terms *motion* and *velocity* are used in this article merely to *illustrate* the meaning of the term "*rate*." It is true that velocity is a rate—the rate of motion. But many other things beside motion have a rate; such as the variation of light, heat, magnetism, force, anything which increases or diminishes by the operation of prescribed law; and the calculus is applicable to all such subjects where the conditions can be expressed analytically.

The idea of *time* in its *absolute sense* is also foreign to the calculus. The term "unit of time" in the definition does not refer to any specific portion of time; it may be great or small; its *value* does not enter into the calculation, and hence this system does not in any wise invade the domain of natural philosophy. All that the *abstract* science of the calculus has to do with *time*, is confined to the simple condition that the suppositive changes in the value of the variable and of its function, which symbolize their rates of change, shall be *simultaneous*. And this is no more than all systems of the calculus require for the *actual* changes which are supposed to be made in the same quantities.

In the *application* of the calculus to Geometry the idea of motion was in some sort introduced; but not, however, in its philosophical sense as having an *absolute* value. Geometrical magnitudes are supposed to be generated by the movement of their elements. Thus a line is generated by the flowing of a point, a surface by a line, and a solid by a surface; and this conception is used to determine the proportion of magnitudes, by comparing the rates at which they are generated instead of comparing the magnitudes themselves with each other. This idea of the generation of magnitudes by means of their elements is not new in mathematics. It is one of the seminal ideas of the Cartesian system; and though in this work it is certainly made more prominent than it has usually been, and more prolific in results, it is not therefore out of place.

DIFFERENTIALS.

(3) *The differential of a variable, or function, is its rate of change or variation, symbolically expressed by the suppositive change that would take place at that rate.*

If the variable is essentially *positive* and *increasing*, or *negative* and *decreasing*, its differential will be essentially *positive*. If it is essentially *negative* and *increasing*, or *positive* and *decreasing*, its differential will be essentially negative. Thus, if we consider a northern latitude positive and a southern negative, a vessel will have a positive rate (or differential) of progress if her northern latitude is increasing or her southern latitude is decreasing; and *vice versa* her rate of progress will be negative.

The *notation* used to designate the rate of variation or differential is the letter d placed before the variable whose rate is required. Thus the differential of x is written dx. If the variable is a component expression such as x^2+ay, the differential would be written $d(x^2+ay)$. Variables are themselves indicated by the last letters of the alphabet.

NOTE.—I use the nomenclature and notation of Leibnitz, not because there is any actual *necessity* for so doing, but because their use has become so general, not only in the system of Leibnitz, but also in other systems, that it seems to have become fixed, without much regard for their original derivation and meaning; and hence a change in that respect would appear like an unnecessary innovation.

The term "fluxion" is more truly significant of the true principles of the science than the term "differential," and the symbol "d" is no better than some other would be: but it is just as good as any, and the use of it involves no inconsistency with the principles on which the system is based.

CONSTANTS.

(4) The other class of quantities which enter into the transcendental analysis is that of *constants*. These are supposed to have a *fixed value*, although it is not always necessary that this value should be known or given. In equations the constant quantities express the conditions of the proposition, and while they are generally supposed to have a given, or, at least, an *assignable value*, there are many cases in which their value must be determined by the solution of an equation just as any unknown quantity in algebra is determined. This, however, does not make them variables; their value is as much fixed as if it were known at first. The solution of an equation is rendered necessary in order to make the immediate conditions conform to some ulterior conditions imposed upon them. Thus the general equations of two circles will determine the curves when the constants are given; but if there is an ulterior condition that they must be tangent to each other, the constants must be made to conform to this condition; which can be done only by an equation from which the necessary values can be obtained. This solution does not *fix* the values of the constants, but only *makes known* those values which were fixed or rendered certain by the conditions to which they were subjected. *A constant then is never in a state of variation.*

FUNCTIONS.

(5) *A function of a variable is any algebraic expression whose value depends on that of the variable.* Thus

$$ax^2 - 2x$$

is a function of x since its value changes with that of x, supposing a to be constant. The expression

$$ax + by$$

is a function of x and y (a and b being constant), for it depends on both x and y for its value; and thus we may have a function of any number of variables.

When the expression does not involve an *equation*, the variables are independent of each other; that is, we may assign to any of them any value whatever without regard to the values assigned to the rest. But an *equation* which contains variables will have at least *one* dependent on the others for its value. Thus in the equation

$$y^2 + bu = ax$$

in which x, y and u are variables, we may give arbitrary values to any two of them, but the value of the third must be determined from the equation. This last is called a function of the others and the *dependent* variable, while the others are called *independent* variables. When the dependent variable stands alone in one member of the equation it is called an *explicit* function of the others, but when combined with the others it is called an *implicit* function. The term function, however, applied to the dependent variable is to be understood as meaning the *representative* of the function, and not literally the function itself.

Functions are commonly divided into two classes, which are distinguished by the manner in which the variables enter into them, and are called "*Algebraic*" and "*Transcendental.*"

Algebraic functions are those in which the variables are subjected only to the operations of addition, subtraction, multiplication, division, and involution or evolution, denoted by constant exponents or indices.

Transcendental functions are those in which the variable is either an exponent, logarithm or trigonometrical line, such as a sine, tangent, etc.

(**6**) The fundamental problem of the calculus is to find the differential, or rate of change, in a function of a variable produced by that of the variable itself.

As the differential or rate of change in a variable is represented by a suppositive change, taking place at a uniform rate, and that of the function arising from it by a corresponding suppositive change, these changes (being uniform from the beginning) will have a constant ratio independent of their value. *Hence the differential of the variable is always a factor of the differential of its function.* The other factor, that is, the ratio between the differential of the variable and that of its function, is called the differential coefficient of the function. Since this ratio is not affected by the value of the suppositive change representing the differential of the variable, this differential is indicated by an indeterminate symbol, and the differential of the function becomes a function of that symbol.

The differential of a function is obtained by a process called *differentiation*, and the differential coefficient is obtained by dividing the differential of the function by that of the variable; so that the differential coefficient of ax^2-bx would be

$$\frac{d(ax^2-bx)}{dx}$$

If we represent the function by u we have

$$u=ax^2-bx$$

and

$$\frac{du}{dx} = \text{differential coefficient, which}$$

we can obtain as soon as we know how to find the differential of ax^2-bx.

(7) If we have an equation containing variables in each member, since the two members are *always* equal, their rates of change are also equal; hence we may differentiate each member as a separate function, and place the results equal to each other. If the equation contains more than one variable, one of them will be dependent and the value of its differential will depend on the values of the other variables and their differentials. Either of the variables may be taken as the dependent one, and it will then represent a function of the rest.

If the differential of a function of two or more variables be taken with reference to *one* only, and then divided by *its* differential, the result will be the differential coefficient for *that* variable, and all the rest must be treated as constants for *that coefficient*, and the function as a function of *that variable*; for *a differential coefficient can exist only between a single variable and its function.*

If we wish to *indicate* a function of any variable, as x, without giving it any particular form, for the purpose of demonstrating some general truth applicable to all forms, we use the expression F (x), which means any function depending on x for its value. If it is a function of two or more variables, the expression is F (x, y), or F (x, y, z), and similarly for a greater number of variables.

SECTION II.

DIFFERENTIATION OF FUNCTIONS.

Proposition I.

(8) *To find the sign with which the differential of the variable must enter that of the function to which it belongs.*

Among the characteristics of quantity as used in the calculus, is that of being essentially positive or negative, according as it lies on one side or the other of the zero point. If, for instance, the value is reckoned toward the negative side it is essentially negative independent of the algebraic sign that may be prefixed to it. Thus the co-sine of 120° is essentially negative, whether it is to be added or subtracted, while the sine of the same angle is essentially positive. These characteristics are independent of, and wholly distinct from, the algebraic signs that may be prefixed to them. Hence when quantities enter into a function as variables we must inquire what will be the effect of these characteristics on the influence which their rate of change will have on that of their function, so that we may give to the differential of the variable that sign which will produce a rate of change in the function in the right direction.

If, then, a variable, whether intrinsically positive or negative, is increasing, *its rate* of change will affect the *rate* of its *function* in the same direction as its *value* affects the *value* of the function; and, hence, having essentially the same character (Art. 3) it must have the same sign.

If it is diminishing, its rate of change will affect that of its function in a direction *contrary* to that in which the *value* of the variable affects the *value* of its function; but having itself a character *contrary* to that of the variable (Art. 3) it must still have the same sign, in order to produce the proper effect.

Let us for instance take the function

$$x-y$$

in which x is a positive and increasing variable. Now if y is increasing its rate will be essentially positive or negative, according as y is itself essentially positive or negative (Art. 3), and must therefore affect the differential or *rate* of the function (to increase or diminish it) in the same direction as y affects its *value*. If y is diminishing, its rate of change will be essentially negative if y is positive, and positive if y is negative (Art. 3), and will therefore affect the *rate* of the function in a direction contrary to that in which y affects its *value*; but being essentially *contrary* in its nature to that of y, it must enter the differential of the function with the same sign in order to produce a contrary effect.

Thus, whether the variable be intrinsically positive or negative, whether it is increasing or diminishing, *the sign prefixed to its differential must in all cases be the same as that prefixed to the variable itself.*

ILLUSTRATION.

If we consider a northern latitude positive and a southern one negative, and there are two vessels, A and B, sailing north of the equator, let the latitude of A be represented by x and that of B by y. Then the difference of their latitudes will be $x-y$. If both are sailing north their rates of progress will be positive, and the rate of change in their difference of latitude will be the difference of their rates of sailing; hence the rate of B, which is the differential of y, will have a minus sign. If B is sailing south, the difference of their latitudes will be still $x-y$; but since the rate of change in this difference is the real sum of their rates of sailing, the rate of B being essentially negative, (Art. 3) must have a minus sign, so that the algebraic difference will produce the real sum.

If B be south of the equator, the difference of their latitudes will be their real sum, but y being now essentially negative must have a minus sign to produce this sum, and it will still be expressed by $x-y$. If B be sailing south (A being still sailing north), the rate of B will be negative (Art. 3), and since the rate of change in the difference of their latitudes is the real sum of their rates of sailing, the rate of B must have a minus sign, so that the algebraic difference will be the real sum. If B be sailing north, its rate will be positive (Art. 3), and since in this case the rate of change in the difference of latitudes will be the real difference of their rates of sailing, the rate of B must still have a minus sign, so that the algebraic difference will correspond to the real difference. Hence, if y enter the function with a minus sign, its differential or rate of change will have a minus sign, whether y is intrinsically positive or negative, or is increasing or diminishing. A similar result would follow if the sign were *plus*, the sign of the differential would be plus.

Proposition II.

(9) *To find the differential of a function consisting of terms connected together by the signs plus and minus.*

That is to say, to find the rate of change in the function arising from the rates of the variables which enter into it.

Every term in an algebraic expression may be considered

as having a *single value*, made up, of course, of the respective values of the quantities that compose it, and their relations to each other, and may, therefore, be expressed by a single letter. If a term contain none but constant quantities the letter representing it will be considered as a constant. If it contain variables, the letter representing it will be considered as a variable having the same rate of change as would arise in the term itself from the rates of change in the variables which enter into it.

It will, therefore, be sufficient to investigate the case of a function in which each one of the terms is represented by a single letter.

Let us suppose some of these terms to be variable and others constant, and the variables to be changing their values at any rate whatever, either uniform or variable, and each one independent of the rest The constants will, of course, have no rate of change, and will, therefore, not affect the rate of change in the function.

The differential of each variable will be the suppositive uniform change that *would take place* in it in a unit of time at the rate existing at the moment of differentiation; and the differential of the function is the uniform change that would take place in it arising from the supposed uniform changes in the variables. Now, if we suppose this symbolic or suppositive change to be made in each variable, the corresponding change in the function will be the algebraic sum of the changes in the variables; and as these are by supposition uniform, the change in the function will be uniform also at the rate at which it commenced, and will, therefore, be the symbol of that rate or the differential of the function.

For example, let us take the function

$$x-y+a-b+z+c$$

in which x, y and z are variables, and a b and c are constants, and represent by dx, dy and dz the differentials or uniform changes that would take place in x, y and z in a unit of time from the moment of differentiation. Let us also suppose these symbolic changes to take place; the function would then become

$$x+dx-y-dy+a-b+z+dz+c$$

(Art. 8) and if from this we subtract the primitive function we have

$$dx-dy+dz$$

Which represents the uniform change in the function arising during the same unit of time from the suppositive uniform changes in the variables. It is, therefore, the symbol representing the corresponding rate of change, or differential of the function. Hence

$$d(x-y+a-b+z+c) = dx-dy+dz$$

or *the differential of a function composed of terms containing independent variables, having any rates of change whatever, the terms being connected together by the signs plus and minus, is the algebraic sum of the differentials of the terms taken separately with the same signs.*

Since each of the terms in the case given may represent a compound term of any form whatever, it is now necessary to examine the method of finding the differential of a single term in every form in which it may occur.

The number of these forms for algebraic terms is limited to seven, as follows:

1. A variable multiplied by a constant.
2. One variable multiplied by another.
3. A variable divided by a constant.
4. A constant divided by a variable.
5. One variable divided by another.
6. A power of a variable.
7. A root of a variable.

These simple forms, or some combinations of them, which can be dissected and operated by the same rules, constitute all that can be assumed by single algebraic terms.

(10) *To find the differential of a variable multiplied by a constant quantity.*

We have seen that

$$d(x+y+z+u) = dx+dy+dz+du$$

If we make these variables each equal to x, we shall have

$$x+y+z+u=4x$$

and

$$dx+dy+dz+du = 4dx$$

hence

$$d(4x)=4dx$$

As the same reasoning will extend to any number of terms, we may make the equation general, and we have

$$d(nx) = ndx$$

That is, the rate of change of n times x is equal to n times the rate of change of x.

Hence, *the differential of a variable, with a constant coefficient, is equal to the differential of the variable multiplied by the coefficient.* In other words, the coefficient of the variable will also be the coefficient of its differential.

EXAMPLES.

Ex. 1. What is the differential of abz? — *Ans.* $abdz$.

Ex. 2. What is the differential of b^2y? — *Ans.* b^2dy.

Ex. 3. What is the differential of $ax+cy$? — *Ans.* $adx+cdy$.

Ex. 4. What is the differential of $x-by$? — *Ans.*

Ex. 5. What is the differential of $(a+b)x$? — *Ans.*

Ex. 6. What is the differential of $(c-d)y$? — *Ans.*

Ex. 7. What is the differential of $ax+by+cz$? — *Ans.*

Ex. 8. What is the differential of b^2u+c^2z? — *Ans.*

Ex. 9. What is the differential of a^2bx+c^2dy? — *Ans.*

Ex. 10. What is the differential of a^2y-b^2x? — *Ans.*

Ex. 11. What is the differential of $b(ay-cx)$? — *Ans.*

Ex. 12. What is the differential of $c^2(bx+az)$? — *Ans.*

LEMMA.

(11) If two variables are increasing at a uniform rate, their rectangle will be increasing at an accelerated rate, but the acceleration will be constant.

Let x and y be increasing uniformly at rates represented by dx and dy. Then dx and dy will be the actual increments of x and y in a unit of time, and at the end of m such units xy will have become

$$(x+mdx)(y+mdy)=xy+mydx+mxdy+m^2dxdy \qquad (1)$$

In one more unit of time we shall have

$$(x+(m+1)dx)(y+(m+1)dy)=xy+(m+1)ydx+(m+1)xdy$$
$$+(m+1)^2dxdy \qquad (2)$$

In still another unit of time we shall have

$$(x+(m+2)dx)(y+(m+2)dy)=xy+(m+2)ydx+(m+2)xdy$$
$$+(m+2)^2dxdy \qquad (3)$$

Subtracting the second member of (1) from that of (2) we have

$$ydx+xdy+(2m+1)dxdy \qquad (4)$$

and subtracting the second member of (2) from that of (3) we have

$$ydx+xdy+(2m+3)dxdy \qquad (5)$$

and subtracting (4) from (5) we have

$$2dxdy \qquad (6)$$

Now the expression (4) is the increment of the product arising from the uniform increments of the variable factors during one unit of time, and expression (5) is the increment of the product during the next equal unit of time arising from the next equal uniform increments of the variables. These increments of the product may therefore be taken to represent its successive mean rates of increase arising from the uniform increase of the variable factors during two equal successive units of time; and the expression (6), which is the difference between these rates will represent their acceleration.

Now since this last is a constant quantity and independent of m, it follows that the acceleration of the mean rates of increase of the product will be constantly the same during every two consecutive units of time, while the factors are increasing at a uniform rate. And since the increase of the variables is continuous and uniform, that of the product will also be continuous and according to a uniform law of some kind; and since for every possible variation in the number and value of the units of time, and in the value of the rates dx and dy the acceleration of the rate of increase of the product is constant for successive periods, it must be so continuously, and equal to twice the product of the rates of increase of the variable factors.

Proposition IV.

(**12**) *To find the differential of the product of two independent variables.*

Let us suppose the two variables A and B to be increasing at any rate whatever, either uniform or variable, and independent of each other. Süppose also that when A has become equal to x, B will have become equal to y, and that dx and dy represent their respective rates of increase at that instant; then they will represent the uniform increments that *would be made* by A and B respectively, in the same unit of time, at these rates; and these suppositive increments are what we have to consider. Suppose again that one-half of each increment be made immediately before A and B become equal to x and y, and the other half afterwards. In the first case the product of A and B, or AB, at the beginning of the increment will be equal to

$$(x-\tfrac{1}{2}dx)(y-\tfrac{1}{2}dy)=xy-\tfrac{1}{2}ydx-\tfrac{1}{2}xdy+\tfrac{1}{4}dxdy$$

and at the end of the unit of time it will be equal to

$$(x+\tfrac{1}{2}dx)(y+\tfrac{1}{2}dy)=xy+\tfrac{1}{2}ydx+\tfrac{1}{2}xdy+\tfrac{1}{4}dxdy$$

Subtracting the first product from the last we have

$$ydx+xdy$$

which represents the difference between the two states of the rectangle AB, or the increment made by it, while the factors are passing from $x - \frac{1}{2}dx$ and $y - \frac{1}{2}dy$ to $x + \frac{1}{2}dx$ and $y + \frac{1}{2}dy$; that is, while the variables are receiving the uniform increments represented by dx and dy, their respective rates of increase at the instant they are equal to x and y, the rectangle is receiving an increment represented by $ydx + xdy$. We are now to show that this increment represents the rate of increase of AB at the moment that dx and dy represent the rates of increase of A and B separately, namely, at the instant they become equal to x and y.

This suppositive increment would not, of course, be made at a uniform rate, but as we have seen (lemma) at a *uniformly increasing rate.* Hence when AB would become xy, and the variables had received half their suppositive increments, the increment of AB would have received half the *increase* of its rate, which would then have become equal to its mean rate for that unit of time. But the mean rate is that by which the increment *would be* made in the same time if it were uniform, and if the increment were made at a uniform rate it would measure the rate of increase of the rectangle. Now the actual increment (represented by $ydx + xdy$) of the rectangle, being made in the same time, as it would be if made uniformly at its mean rate, existing when A and B were equal to x and y, *is the true measure of that rate;* that is, of the rate of increase of AB at that instant; or of xy if we consider x and y as the variables. Hence *the differential of the product of two variables is equal to the sum of the products arising from the multiplication of each variable by the differential of the other.*

NOTE.— This proposition being the *key* to the whole subject of the differential calculus, should be carefully studied and well understood. The result of this proposition might have been surmised by considering that a product of two variables is subject to two independent causes which produce its rate of change. If x has a certain rate of increase, that of the product will, from *that* cause, be y times that rate; and if y have a certain rate of increase, that of the product will from *that* cause be x times that rate; and the total effect of both causes will be the sum of the partial effects arising from each cause independent of the other; that is, the entire rate of increase of xy, is y times that of x plus x times that of y. This, however, is not *mathematical proof*—it only makes the result *probable*.

This proposition may be illustrated geometrically thus:

Let x be represented by the line AB (Fig. 1), and y by the line AC; then the product xy will be represented by the rectangle $ABDC$. Suppose x and y to be each increasing in such a manner that when x has become equal to AB, and is then increasing at a rate that, if continued, would produce the increment BB' in a unit of time, y will have become equal to AC and be increasing at a rate that would produce the increment CC' in the same unit of time. Then BB' will be dx and CC' will be dy, for they will be the true symbols of the *rates* of increase of x and y. Now the uniform increment, BB', of x will produce the uniform increment $B\,D\,E\,B'$ of the rectangle, which will therefore represent *its* rate of increase arising from that of x. In like manner $C\,C'\,F\,D$ will represent the rate of increase arising from that of y. Hence the symbol of the entire rate of increase of the rectangle arising from the rates of increase of the sides, will be the sum of the two rectangles $B\,B'\,E\,D$ and $C\,C'\,F\,D$ or $ydx+xdy$.

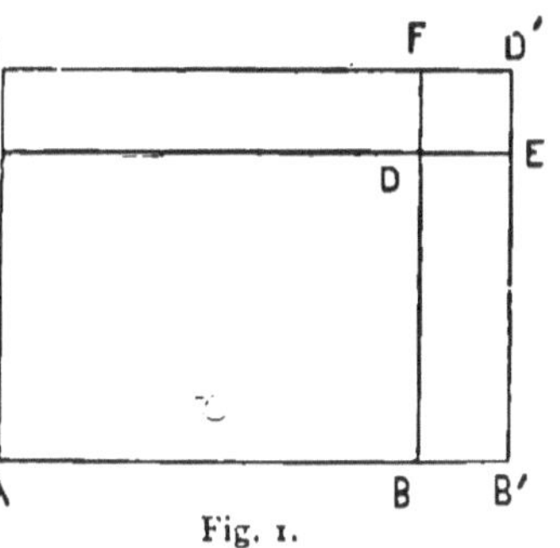

Fig. 1.

It must be remembered that the increments DE and DF are *not actual* increments given to the sides BD and CD, but suppositive increments, or *symbols*, representing their rates of increase; for in order that the suppositive increment of the rectangle may be uniform, the sides BD and CD must remain constant. Hence the two rectangles BB'ED and CC'FD are not actual increments of the rectangle ABDC, but suppositive increments, which are *symbols* showing what *would be* its increment if the rate were to remain constant from that point in the value of x and y for a unit of time, and are, therefore, the measure of that rate.

NOTE.—The increment of the rectangle arising from the *actual* increments of x and y would include the small rectangle DED′F or $dx \cdot dy$; and hence it would be too great to represent the required rate by so much; and it is to get rid of this surplus that the advocates of the infinitessimal theory, who take the actual increment to represent the rate, reduce it to an infinitessimal, which they claim may be neglected with impunity. But we see the true reason for throwing it out of the expression for the rate is, *not* because of its insignificance, but because, being produced by the actual increments of x and y *after* they have passed *beyond* the values of AB and AC, it has no connection with the rate at which the rectangle was increasing at the moment x and y were *equal* to those values.

PROPOSITION V.

(13) *To find the differential of the product of any number of variable factors.*

Let us first take the product of three variables as

$$xyz$$

If we represent the product xy by u we shall have

$$xyz = uz$$

and (Art. 7 and 12)

$$d(xyz) = d(uz) = zdu + udz$$

Replacing u by its value we have

$$d(xyz) = zd(xy) + xydz$$

or since

$$d(xy) = ydx + xdy$$

we shall have

$$d(xyz) = zydx + zxdy + xydz$$

If we take the product of four variables as

$$xyzu$$

and make $xy = r$ and $zu = s$ we have

$$xyzu = rs$$

and

$$d(xyzu) = d(rs) = rds + sdr$$

Replacing r and s by their values we have

$$d(xyzu)=xyd(zu)+zud(xy)$$

substituting the values of $d(zu)$ and $d(xy)$ we have

$$d(xyzu)=xyzdu+xyudz+uzxdy+uzydx$$

These examples may be carried to any extent, but are enough to show the *law* which governs the differential of a product, which law may be thus stated: The rate of change in a product arising from the rate of any one factor is equal to the product of all the other factors multiplied by that rate; and the total rate of change in the product is the sum of all the partial rates arising from the rates of the factors taken separately. In other words, the total effect arising from all the causes acting together is equal to the sum of the partial effects arising from each cause acting separately.

Hence, *the differential of the product of any number of variable factors is equal to the sum of the products arising from multiplying the differential of each variable by the product of all the other variables.*

EXAMPLES.

Ex. 1. What is the differential of ayz? —

Ans. $aydz+azdy$

Ex. 2. What is the differential of $4\,bxy$? —

Ans. $4\,b(xdy+ydx)$

Ex. 3. What is the differential of $ax+byz$? —

Ans. $adx+bydz+bzdy$

Ex. 4. What is the differential of $xy-uz$? — *Ans.*

Ex. 5. What is the differential of $3ax-2xy$? — *Ans.*

Ex. 6. What is the differential of $2ay+3u$? — *Ans.*

Ex. 7. What is the differential of $4abxyz$? — *Ans.*

Ex. 8. What is the differential of $bcu-a^2-zy$? — *Ans.*

Ex. 9. What is the differential of $4axy-b^2+cu$? — *Ans.*

Ex. 10. What is the differential of $7a^2-2bu+cy$? — *Ans.*

PROPOSITION VI.

(14) *To find the differential of a fraction.*

CASE 1.

Let the fraction be

$$\frac{x}{n}$$

in which the variable is divided by a constant. Make

$$\frac{x}{n}=y$$

then

$$x=ny$$

and (Art. 10)

$$dx=ndy$$

hence

$$dy=d\left(\frac{x}{n}\right)=\frac{dx}{n}$$

that is,

The differential of a fraction having a variable numerator and a constant denominator is equal to the differential of the numerator divided by the denominator.

CASE 2.

Let the fraction be

$$\frac{n}{x}$$

in which the denominator is variable and the numerator constant. Make

$$\frac{n}{x}=y$$

then

$$xy=n$$

and (Art. 12 and Art. 9)

$$d(xy)=ydx+xdy=dn=o$$

hence

$$dy=-\frac{ydx}{x}$$

Replacing y by its value we have

$$d\left(\frac{n}{x}\right)=-\frac{\frac{n}{x}dx}{x}=-\frac{ndx}{x^2}$$

that is,

The differential of a constant divided by a variable is equal to minus the numerator into the differential of the denominator, divided by the square of the denominator.

CASE 3.

Let the fraction be

$$\frac{x}{y}$$

in which both terms are variable, Make

$$\frac{x}{y}=u$$

then

$$x=uy$$

and

$$dx=d(uy)=ydu+udy$$

hence

$$du=\frac{dx-udy}{y}$$

Replacing u by its value we have

$$d\left(\frac{x}{y}\right)=\frac{dx-\frac{x}{y}dy}{y}=\frac{ydx-xdy}{y^2}$$

that is,

The differential of a fraction of which both terms are variable, is equal to the differential of the numerator multiplied by the denominator, minus the differential of the denominator multiplied by the numerator, and this difference divided by the square of the denominator.

If the variables in any of these cases should be functions of other variables, we can represent them by single letters and replace them in the formula by their values

Thus, suppose the given fraction to be

$$\frac{u-x}{vy}$$

make

$$u-x=s \text{ and } vy=r$$

then

$$d\left(\frac{u-x}{vy}\right)=d\left(\frac{s}{r}\right)=\frac{rds-sdr}{r^2}$$

Replacing the values of s and r we have

$$d\left(\frac{u-x}{vy}\right)=\frac{vyd(u-x)-(u-x)d(vy)}{v^2y^2}$$

which being expanded becomes

$$d\left(\frac{u-x}{vy}\right)=\frac{vydu-vydx-uvdy-uydv+vxdy+xydv}{v^2y^2}$$

EXAMPLES.

Ex. 1. What is the differential of $ax+\frac{y}{b}$?

Ans. $adx+\frac{dy}{b}$

Ex. 2. What is the differential of $cxy+\frac{a}{u}$?

Ans. $cxdy+cydx-\frac{adu}{u^2}$

Ex. 3. What is the differential of $\frac{a}{x}-\frac{c}{y}$? *Ans.*

Ex. 4. What is the differential of $\frac{a-b}{x-c}$? *Ans.*

Ex. 5. What is the differential of $5x+\frac{2x}{c-y}$? *Ans.*

Ex. 6. What is the differential of $3(a-x)-\frac{y-c}{u-d}$? *Ans.*

Ex. 7. What is the differential of $2ax+x(y-c)$? *Ans.*

Ex. 8. What is the differential of $4(by-x)(u-c)$? *Ans.*

Ex. 9. What is the differential of $\frac{a-x}{b-y}\times\frac{d+u}{v}$? *Ans.*

Ex. 10. What is the differential of $(a-v)(xy-z)$? *Ans.*

Ex. 11. What is the differential of $6ab-3xy\,(u-c)$? *Ans.*

Ex. 12. What is the differential of $(z-y)(a-x)+\frac{c-x}{y-z}$? *Ans.*

PROPOSITION V.I.

(15) *To find the differential of any power of a variable.*

We have seen (Art. 13) that the differential of the product of any number of variables is equal to the sum of the products arising from multiplying the differential of each variable by the product of the rest. If there are n variable factors in the given product, there will be n products in the differential, and the coefficient of the differential of each variable will be the product of $(n-1)$ variables. If now all the variables become equal, we may represent each one by x, and the product will become x^n, while each of the products composing the differential will become $x^{n-1}dx$; and as there are n of these products, the sum of them will be $nx^{n-1}dx$, hence we have

$$d\,(x^n)=nx^{n-1}dx$$

Hence, *the differential of a variable raised to a power is equal to the variable raised to the same power less one, and multiplied by the exponent of the power and the differential of the variable.*

If, for example, we take

$$d\,(xyzu)=xyzdu+xyudz+xzudy+yzudx$$

and suppose all the variables to become equal, we shall have

$$dx^4=4x^3dx$$

If we represent x^n by y we shall have

$$dy=nx^{n-1}dx$$

and

$$\frac{dy}{dx}=nx^{n-1}$$

which is the *differential coefficient* of that function of x.

(16) The rule here given, for finding the differential of the power of a variable, holds good for all values of n, whether integral or fractional, positive or negative.

Case 1. Let n be negative and represented by $-m$, then

$$x^n=x^{-m}=\frac{1}{x^m}$$

and (Art. 14)

$$d\left(\frac{1}{x^m}\right)=-\frac{d(x^m)}{x^{2m}}=-\frac{mx^{m-1}dx}{x^{2m}}$$

Dividing both terms of this fraction by x^{m-1} we have

$$-\frac{mdx}{x^{m+1}}=-mx^{-m-1}dx$$

in accordance with the rule.

Case 2. Let n be equal to $\frac{1}{m}$ then

$$x^n=x^{\frac{1}{m}}$$

Representing $x^{\frac{1}{m}}$ by y we have

$$x=y^m$$

and

$$dx=my^{m-1}dy$$

or

$$dy=\frac{dx}{my^{m-1}}$$

Replacing y by its value we have

$$d\left(x^{\frac{1}{m}}\right)=\frac{dx}{m\left(x^{\frac{1}{m}}\right)^{m-1}}=\frac{dx}{mx^{1-\frac{1}{m}}}=\frac{1}{m}x^{\frac{1}{m}-1}dx$$

Case 3. Let n be equal to $\frac{r}{m}$ and represent $x^{\frac{r}{m}}$ by y and we have

$$y=x^{\frac{r}{m}} \text{ or } y^m=x^r$$

and

$$my^{m-1}dy=rx^{r-1}dx$$

whence

$$dy=\frac{rx^{r-1}dx}{my^{m-1}}$$

Replacing y by its value, we have

$$d\left(x^{\frac{r}{m}}\right)=\frac{r}{m}\times\frac{x^{r-1}dx}{\left(x^{\frac{r}{m}}\right)^{m-1}}=\frac{r}{m}x^{r-1-\frac{r}{m}(m-1)}dx=\frac{r}{m}x^{\frac{r}{m}-1}dx$$

hence the rule is true in all cases.

PROPOSITION VIII.

(17) *To find the differential of any root of a variable.*

Let us take the function $\sqrt[m]{x}$ and make it equal to y, then

$$y^m = x$$

and

$$my^{m-1}dy = dx$$

whence

$$dy = \frac{dx}{my^{m-1}}$$

Replacing y by its value we have

$$d\sqrt[m]{x} = \frac{dx}{m\sqrt[m]{x^{m-1}}}$$

That is, *the differential of any root of a variable is equal to the differential of the variable divided by the index of the root into the same root of the variable raised to a power denoted by the index of the root less one.*

Hence

$$d\sqrt{x} = \frac{dx}{2\sqrt{x}}$$

These results could of course be obtained under the preceding rule, by giving to the variable a fractional exponent. But this is not always convenient.

(18) In all the cases which have been explained under these eight propositions, the single letters that have been used may represent functions of other variables, in which case the operation must be continued by substituting the functions for the letters representing them and performing upon *them* the operations indicated, which can be done by the rules already given; for the terms of all algebraic functions must ultimately take some one of the forms that we have discussed.

Thus, if we have the function

$$\frac{x^2-\sqrt{y}}{a-x}$$

we may represent the numerator by u and the denominator by v we shall then have

$$\frac{x^2-\sqrt{y}}{a-x}=\frac{u}{v}$$

and

$$d\left(\frac{x^2-\sqrt{y}}{a-x}\right)=\frac{vdu-udv}{v^2}$$

Replacing u and v by their values we have

$$d\left(\frac{x^2-\sqrt{y}}{a-x}\right)=\frac{(a-x)d(x^2-\sqrt{y})-(x^2-\sqrt{y})d(a-x)}{(a-x)^2}$$

and performing the operations indicated we have

$$d\left(\frac{x^2-\sqrt{y}}{a-x}\right)=\frac{(a-x)\left(2xdx-\frac{dy}{2\sqrt{y}}\right)+(x^2-\sqrt{y})dx}{(a-x)^2}$$

which completes the differentiation.

PROPOSITION IX.

(**19**) To find the differential of a function with respect to the independent variable when it enters the given function by means of another function of itself.

Let there be a function of y in which y represents a function of x. The proposition is to find the differential co-efficient of the given function with respect to x. Representing the given function by u we have

$$u=F(y) \text{ and } y=F(x)$$

In order to find the relation between du and dx, it might be supposed necessary to eliminate y from the equations and obtain one directly between u and x, to which the ordinary process of differentiation could be applied. But this is not necessary; for if we differentiate these equations as they are we have

$$du=F'(y)dy \text{ and } dy=F'(x)dx$$

in which $F'(y)$ and $F'(x)$ represent the differential coefficient of u with respect to y and of y with respect to x. From the first of these we obtain

$$dy=\frac{du}{F'(y)}$$

and placing this value of dy equal to the other we have

$$\frac{du}{F'(y)}=F'(x)dx \text{ or } \frac{du}{dx}=F'(y) \,.\, F'(x)$$

That is, the differential coefficient of u with respect to x is equal to that of u with respect to y multiplied by that of y with respect to x. Hence

When the variable of a given function represents the function of an independent variable, then *the differential coefficient with respect to the independent variable is equal to the product of the differential coefficient of the function with respect to the given variable, multiplied by the differential coefficient of the given variable with respect to the independent variable.*

Thus if we have the function y^2-ay in which $y=2a-x^2$, representing the given function by u, we have

$$\frac{du}{dy}=2y-a \text{ and } \frac{dy}{dx}=-2x$$

whence

$$\frac{du}{dx}=2ax-4xy=4x^3-6ax$$

EXAMPLES.

Find the differentials of the following functions in which the variables are independent.

Ex. 1. a^2x^2+z *Ans.* $2a^2xdx+dz$

Ex. 2. bx^2-y^3+a *Ans.* $2bxdx-3y^2dy$

Ex. 3. ax^2-bx^3+x *Ans.*

Ex. 4. $(c+d)(y^2-x^2)$ *Ans.*

Ex. 5. $5x^5-2ay-b^2$ *Ans.*

Ex. 6. x^n-x^3+4b *Ans.*

Ex. 7. ax^3-3bx *Ans.*

Ex. 8. $(x^2+a)(x-a)$ *Ans.*

Ex. 9. $x^2y^2-z^2$ *Ans.*

Ex. 10. $ax^2(x^3+a)$ *Ans.*

Ex. 11. $\dfrac{a}{b-2y^2}$ *Ans.*

Ex. 12. $\sqrt{a^2-x^2}$ *Ans.*

Ex. 13. $\sqrt{2ax+x^2}$ *Ans.*

Ex. 14. $\dfrac{1}{\sqrt{1-x^2}}$ *Ans.*

Ex. 15. $\dfrac{x}{x+\sqrt{1-x^2}}$ *Ans.*

Ex. 16. $(a+\sqrt{x})^3$ *Ans.*

Ex. 17. $\dfrac{a^2-x^2}{a^4+a^2x^2+x^4}$ *Ans.*

Ex. 18. $\dfrac{x^n}{1+x^n}$ *Ans.*

Ex. 19. $\dfrac{1+x^2}{1-x^2}$ *Ans.*

Ex. 20. $\dfrac{\sqrt{1+x}-\sqrt{1-x}}{\sqrt{1+x}+\sqrt{1-x}}$ *Ans.*

Ex. 21. $(a+bx^n)^m$ *Ans.*

Ex. 22. $\sqrt{ax^3}$ *Ans.*

Ex. 23. $x^{\frac{1}{2}}y^{-\frac{1}{2}}$ *Ans.*

Ex. 24. $mx^{-1}+(x^2y)^{\frac{3}{2}}$ *Ans.*

Ex. 25. $x^2y^{\frac{1}{2}}-z$ *Ans.*

Ex. 26. $a+bcx-3y^{\frac{1}{4}}$ *Ans.*

Ex. 27. $x^{-3}+y^{-1}+z^3$ *Ans.*

Ex. 28 $(ax-y)^{\frac{2}{5}}$ *Ans.*

Ex. 29. $(b-c)(x-y)^{\frac{1}{2}}$ *Ans.*

Ex. 30. $\sqrt{x^2+a\sqrt{x}}$ *Ans.*

Ex. 31. A person is walking towards the foot of a tower, on a horizontal plain, at the rate of 5 miles an hour; at what rate is he approaching the top, which is 60 feet high, when he is 80 feet from the bottom?

Let the height of the tower be a, the distance of the person from the foot of it be x, and the distance from the top be y; then

$$y^2=a^2+x^2$$

and

$$ydy=xdx$$

whence

$$dy=\frac{xdx}{y}=\frac{80\,ft.\times5\,m.}{100}=4\,m.$$

Ans. 4 *miles per hour.*

Ex. 32. Two ships start from the same point and sail, one north at the rate of 6 miles per hour, and the other east at the rate of 8 miles per hour; at what rate are the ships leaving each other at the end of two hours?

Ans. 10 *miles per hour.*

Ex. 33. Two vessels sail directly south from two points on the equator 40 miles apart; one sails at the rate of 5 miles per hour, and the other at the rate of 10 miles per hour; how far will they be apart at the end of 6 hours, and at what rate will they be separating from each other, supposing the meridians to be parallel? *Ans.* 3 *miles per hour.*

Ex. 34. A ship sails directly south at the rate of 10 miles per hour, and another ship sailing due west crosses her track two hours after she has passed the point of crossing, at the rate of 8 miles per hour; at what rate are they leaving each other one hour afterwards?

Ans. 11.74 *miles per hour nearly.*

Ex. 35. The vessels sailing as in the last case, how will

the distance between them be changing, and at what rate, one hour before the second crosses the track of the first?

At that time the first vessel will be 10 miles from the point of crossing, and the second 8 miles. Calling the distance of the first x, and the second y, the distance between them is $\sqrt{x^2+y^2}$ which we will call u, then

$$u=\sqrt{x^2+y^2}$$

and

$$du=\frac{xdx+ydy}{\sqrt{x^2+y^2}}=\frac{100-64}{\sqrt{100+64}}$$

We make 64 in the numerator negative because y is a positive decreasing function, and its differential is therefore intrinsically negative.

From the above equation we find that the vessels are separating at the rate of 2.812 miles per hour nearly.

Ex. 36. The height of an equilateral triangle is 24 inches, and is increasing at the rate of two inches per day; how fast is the area of the triangle increasing?

Ans. $32\sqrt{3}$ *square inches.*

Ex. 37. The diameter of a cylinder is 2 feet, and is increasing at the rate of 1 inch per day, while the height is 4 feet and decreasing at the rate of 2 inches per day; how is the volume changing? and how the convex surface?

Ans. The volume is increasing at the rate of 288 π cubic inches per day. The area of the convex surface is not changing.

SECTION III.

SUCCESSIVE DIFFERENTIALS.

(20) In considering the differential of a function hitherto, we have regarded it as immaterial whether that of the independent variable was itself variable or uniform. In considering the rate of change in the *differential* of the function, it will be most convenient to consider that of the independent variable as *constant;* and this we have a right to do, since, the variable being independent, its rate is always assignable.

If we have the product of two or more independent variables, whether they are alike as x^3, or different as $x.y.z$, the value of the rate of change depends not only on that of each independent variable, but also on the absolute value of all the variables. Thus the value of $d(x^3)$, or $3x^2dx$, depends not only on that of dx, but also on the absolute value of x^2, and is greater or less as x^2 is greater or less. So that $3x^2dx$, which is the rate of change of x^3, has its own rate of change or differential.

To find this second differential we must treat the first as an original function; and as x is supposed to change uniformly, dx will be regarded as a constant. Now the function $3x^2dx$ may take the form $3dx.x^2$, and

$$d(3dx.x^2)=3dx.d(x^2)=3dx.2xdx=6xdx^2$$

This is called the second differential of x^3, being the differential of the differential. This order is indicated by placing the figure 2 as a sort of exponent to the letter d, thus $d^2(x^3)$ is the symbol for the second differential of x^3, and hence

$$d^2(x^3)=6xdx^2$$

If the function is at all complicated, and, especially, if we desire to indicate a differential coefficient, it is much more convenient to represent it by a single letter; in which case the letter itself is, for the sake of brevity, called the function.

So that if we represent x^3 by u we shall have

$$u=x^3 \text{ and } d^2u=6xdx^2 \qquad (1)$$

Since the second member of this last equation still contains the letter x, it will still have a rate of change which may be found by considering $6dx^2$ as the constant coefficient of x, and differentiating we have (Art. 10)

$$d^3u=6dx^2.dx=6dx^3 \qquad (2)$$

The expressions dx^2 and dx^3 are to be understood as indicating, *not* the differentials of x^2 and x^3, but the square and cube of dx.

The figure placed as an exponent to the letter d indicates the *order* of the differential; and the differential of any order above the first is the rate of change in the differential of the previous order. The differentiation of the differential can take place only while the latter represents what is still a function of the independent variable. The differential of an independent variable, being, as we have stated, supposed to be constant, can have no differential.

If we divide equations (1) and (2) respectively by dx^2 and dx^3, we have

$$\frac{d^2u}{dx^2}=6x \text{ and } \frac{d^3u}{dx^3}=6$$

in which the second members are the second and third differential coefficients of the function x^3.

(21) To illustrate the principle of successive differentiation, let us suppose *A B D C F E G* (Fig. 2) to represent a cube, of which the side *A B* is an increasing variable represented by x. Let the suppositive increments *Dd*, *Dd'* and *Dd''* represent the three equal rates of increase of the three x's whose product is the given cube; we are to find what will be the corresponding rate of increase of the cube itself. That is, *Dd* being equal to dx, what is the value of $d(x^3)$.

At the moment the sides of the cube become equal to x, the cube tends to expand by the movement of the three faces *DA*, *DF* and *DG* outward in the directions *Dd*, *Dd'* and *Dd''*, each face continuing parallel to itself, and thus increasing the cube. But in order that these increments may represent the *rate* at which the cube was increasing when they began, they must be made *uniformly* at the same rate. Hence the areas of the faces must remain constant, and they must move at the same rate as the increments *Dd*, *Dd'* and *Dd''* are described; the movement being controlled by those increments. Thus the three solids *Da*, *Df* and *Dg*, generated by the flowing out of the surfaces *DA*, *DF* and *DG*, with an unchanging area, at the same rate as when their sides became equal to x, form the increment

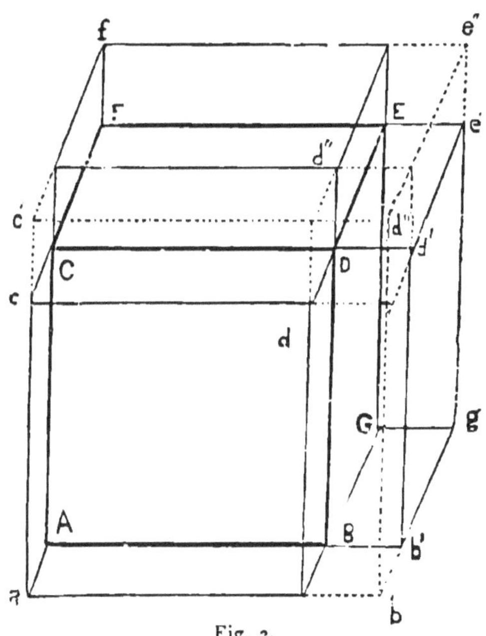

Fig. 2.

that *would take place* in the cube in a unit of time, at the rate at which it was increasing when its side, or edge, became equal to x; and hence they form the true symbolic increment which represents the rate of increase or differential of x^3. Now each of the solids thus formed is equal to x^2dx, and hence

$$d(x^3)=3x^2dx$$

But if the cube, when its edge is equal to x, is in a state of increase, the faces DA, DF and DG have other tendencies besides that of flowing directly outward. *They also* tend to expand in the direction of their own sides, and this tendency is quite distinct from the other. Let us examine and measure it. The tendency of the face DF to expand, arising from that of the cube itself, would be by the flowing out of the sides DC and DE in the directions Dd and Dd', and remaining parallel to themselves. The *rate* of increase of the face DF would be measured by the areas described by these flowing lines in a unit of time, at the same rate as when they became equal to x, their lengths being constant; that is, by the areas Dc and Dc'. But the face DF of the cube is the base of the solid Df, and the tendency of the base to expand imparts a like tendency to the solid, and the rate at which the solid tends to expand is such that while the base would increase by the rectangles Dc and Dc', the solid would increase by the solids Dc' and Dc'', which therefore represent, symbolically, the rate of increase of Df. Now Dc' and Dc'' are each equal to xdx^2, and hence the rate of increase of Df is equal to $2xdx^2$. But the differential of the cube is represented by three such solids as Df and the rate of the increase of the whole, or the differential of the differential of x^3 is $6xdx^2$.

Again the solid Dc' or xdx^2 tends to increase in the direction Dd' at such a rate as would generate the suppositive increment Dd''', equal to dx^3 in the same unit of time and in a uniform manner. Hence $d(xdx^2)$ is equal to dx^3, and, therefore, $d(6xdx^2)$ is equal to $6dx^3$.

(22) It must always be remembered that the solids represented in the figure are *not* the *actual* increments of the cube, but the symbols which represent its *rate of increase* and the successive rates of that rate at the instant that x is equal to AB; that is, they are the increments that *would take place* in the cube. and in the increments themselves if made uniformly. The *law* which governs the increase of the cube, contains within itself not only the rate at which the cube is, at any instant increasing, but also the law of change in that rate, and the law to which that law is subject; and these symbols represent the development of that law which was actually operating at the instant the cube attained the value of AB^3 or x^3, and before any farther increase had taken place.

(23) We may learn from this demonstration the method by which the *actual* increment of a power may be developed. By dissecting the figure (Fig. 2) and noticing the parts of which the increased cube is composed, we find, *first*, the original cube or x^3; *second*, the three solids Da, Df and Dg or $3x^2dx$; *third*, the three solids Dc', Dc'' and Db which represent *half* the rate of increase of $3x^2dx$ or $\frac{6xdx^2}{2}$; and *fourth*, the solid or small cube Da''', which represents one-third of the rate of increase of $\frac{6xdx^3}{2}$ or $\frac{6dx^3}{2 \cdot 3}$; and these make up the volume of the cube after being increased by the addition arising from the increment dx to the side AB. But these increments are suppositive, and are used merely as *symbols* to show the successive rates of increase. all of which exist in the function x^3 before any increment actually takes place.

If we divide $3x^2dx$ by dx we shall have the first differential coefficient of x^3. If we divide $\frac{6x dx^2}{2}$ by dx^2 we shall have *half* the second differential coefficient of x^3; and if we divide $\frac{6 dx^3}{2 \cdot 3}$ by dx^3 we shall have *one-sixth* of the third differential coefficient of x^3; and these results, viz.: $3x^2$, $\frac{6x}{2}$ and $\frac{6}{2 \cdot 3}$ are the coefficients by which the successive powers of dx must be multiplied in order to make up the parts composing the suppositive increment of x^3. Now since these partial increments taken together with the original cube form also a complete cube, if we make dx a *real* increment and multiply its successive powers by these same coefficients, we shall have an *actual* increment of the cube, and the original x^3 will have become $(x+dx)^3$.

It must not be forgotten that these differential coefficients are true of the cube *before* the increment takes place, and when dx is equal to zero

For convenience we will designate the variable cube by u, and in order to *mark* the point where the differential is to be taken we represent its variable edge by $(h+x)$ in which h represents the side AB, or that particular value of the variable where the differential is to be taken, while x will represent its variable increment and take the place of dx. Then the cube AE or $\overline{AB}^3$ will be represented by u or $(h+x)^3$ at the time when the variable u equals h or $x=o$.

This being premised we take the differential coefficients already found, namely, $3x^2$, $\frac{6x}{2}$ and $\frac{6}{2 \cdot 3}$, and substitute $(h+x)$ for x (reducing x to zero at the same time) and x for dx in the other factors; then $3x^2$ becomes the first differential coefficient of u or $(h+x)^3$, that is $\frac{du}{dx}=3(h+x)^2$ with $x=o$, or $3h^2$; $\frac{6x}{2}$ becomes $\frac{6(h+x)}{2}$ with $x=o$; that is $3h$, or half the second differential coefficient of u with $x=o$; and $\frac{6}{2 \cdot 3}$

becomes one-sixth of the third differential coefficient of u or $\frac{d^3u}{6dx^3}=1$.

Hence, indicating by a vinculum that x has been made equal to zero, we have for the three coefficients $3x^2$, $\frac{6x}{2}$ and $\frac{6}{2\,.\,3}$, which were true of the cube before any increment was made,

$$\left(\frac{du}{dx}\right)\ \left(\frac{d^2u}{2dx^2}\right)\ \text{and}\ \left(\frac{d^3u}{2\,.\,3.\,dx^3}\right)$$

also true of the cube at the same time; and the different parts of the cube increased will be represented as follows:

The cube AE by (u) or h^3.

The three solids Da, Df Dg by $\left(\frac{du}{dx}\right)x$ or $3h^2x$.

The three solids Dc', Dc'' and Db by $\left(\frac{d^2u}{2dx^2}\right)x^2$ or $3hx^2$.

The solid Dd''' by $\left(\frac{d^3u}{2\,.\,3\,.\,dx^3}\right)x^3$ or x^3,

and these make up the value of $(h+x)^3$, hence

$$u \text{ or } (h+x)^3=(u)+\left(\frac{du}{dx}\right)x+\left(\frac{d^2u}{2dx^2}\right)x^2+\left(\frac{d^3u}{2.\,3.\,dx^3}\right)x^3$$

or

$$(h+x)^3=h^3+3h^2x+3hx^2+x^3$$

This illustrates to some extent the law which connects together the parts which go to make up the change in the function of a variable arising from that of the variable itself. A more complete and general demonstration of this law is contained in the following theorem.

Maclaurin's Theorem.

(24) A function of a single variable may often be expanded into a series by the following method.

Representing the function by u and the variable by x we shall have

$$u=F(x)$$

When this function can be developed, the only quantities that can appear in the development, besides the powers of x, will be constant terms and constant coefficients of those powers. Hence the developed function may be put into the following form:

$$u=A+Bx+Cx^2+Dx^3+Ex^4+\text{ etc.} \qquad (1)$$

in which A, B, C, D, E, etc., are independent of x. The problem is to find the value of this constant term A, and the values of the constant coefficients B, C, D, E, etc. For this purpose we differentiate equation (1) successively, and divide each result by dx, the successive differential coefficients will then be

$$\frac{du}{dx}=B+2Cx+3Dx^2+4Ex^3+\text{ etc.} \qquad (2)$$

$$\frac{d^2u}{dx^2}=2C+2\cdot3\cdot Dx+3\cdot4\cdot Ex^2+\text{ etc.} \qquad (3)$$

$$\frac{d^3u}{dx^3}=2\cdot3\cdot D+2\cdot3\cdot4\cdot Ex+\text{ etc.} \qquad (4)$$

Since x is an independent variable, these equations are true for all values of x, and, of course, when $x=o$.

Reducing x to zero in equation (1), A becomes equal to the original function with x reduced to zero. We will represent that state of the function by (u); and also indicate by brackets around the differential coefficients that $x=o$ in *their* values also. Then from equations (2), (3), (4), etc., we have

$$\mathrm{B}=\left(\frac{du}{dx}\right),\quad 2C=\left(\frac{d^2u}{dx^2}\right),\quad 2\cdot3\cdot D=\left(\frac{d^3u}{dx^3}\right)$$

and so on to the end of the development, if it can be completed; if not, then in an unlimited series.

Substituting these values of A, B, C, D, etc., in equation (1) we have

$$u=(u)+\left(\frac{du}{dx}\right)x+\left(\frac{d^2u}{dx^2}\right)\frac{x^2}{2}+\left(\frac{d^3u}{dx^3}\right)\frac{x^3}{2.3.}+\text{etc.}$$

which is Maclaurin's theorem.

EXAMPLES.

Ex. 1. Expand $(a+x)^n$ into a series.

Represent $(a+x)^n$ by u and we shall have

$$u=(a+x)^n=A+Bx+Cx^2+Dx^3+\text{ etc.} \qquad (1)$$

Differentiating we have

$$\frac{du}{dx}=n(a+x)^{n-1}$$

$$\frac{d^2u}{dx^2}=n(n-1)\ (a+x)^{n-2}$$

$$\frac{d^3u}{dx^3}=n(n-1)(n-2)(a+x)^{n-3}$$

$$\frac{d^4u}{dx^4}=n(n-1)(n-2)(n-3)(a+x)^{n-4}$$

from which, when $x=o$, we have

$$A=a^n$$

$$B=na^{n-1}$$

$$C=\frac{n(n-1)}{2}a^{n-2}$$

$$D=\frac{n(n-1)(n-2)}{2\cdot 3}a^{n-3}$$

$$E=\frac{n(n-1)(n-2)(n-3)a^{n-4}}{2\cdot 3\cdot 4}$$

Substituting these values in equation (1) we have

$$u=(a+x)^n=a^n+na^{n-1}x+\frac{n(n-1)}{2}a^{n-2}x^2+\frac{n(n-1)(n-2)}{2.3}a^{n-3}x^3$$

and so on; the same result as by the binomial theorem.

Ex. 2. Develop $\frac{1}{a+x}$ into a series.

We have by differentiation

$$\frac{du}{dx}=-\frac{1}{(a+x)^2}$$
$$\frac{d^2u}{dx^2}=\frac{2}{(a+x)^3}$$
$$\frac{d^3u}{dx^3}=-\frac{2\cdot 3}{(a+x)^4}$$
and so on.

Making $x=o$ in the values of u and the differential coefficients we have

$$(u)=\frac{1}{a},\quad \left(\frac{du}{dx}\right)=-\frac{1}{a^2},\quad \left(\frac{d^2u}{dx^2}\right)=\frac{2}{a^3},\quad \left(\frac{d^3u}{dx^3}\right)=-\frac{2\cdot 3}{a^4}$$

Substituting these values in Maclaurin's formula we have

$$u=\frac{1}{a+x}=\frac{1}{a}-\frac{x}{a^2}+\frac{x^2}{a^3}-\frac{x^3}{a^4}+\text{ etc.}$$

Ex. 3. Develop the function $\frac{1}{1-x}$ into a series.

Ans. $1+x+x^2+x^3+x^4+$ etc.

Ex. 4. Develop the function $\sqrt{a+x}$ into a series.

Ans. $a^{\frac{1}{2}}+\frac{1}{2}a^{-\frac{1}{2}}x-\frac{1}{2.4}a^{-\frac{3}{2}}x^2+\frac{1.3}{2.4.6}a^{-\frac{5}{2}}x^3-$ etc.

Ex. 5. Develop $\frac{1}{(a+x)^2}$ into a series. *Ans.*

Ex. 6. Develop $\sqrt[3]{(a+x)^2}$ into a series. *Ans.*

Ex. 7. Develop $\sqrt{a^2+bx}$ into a series. *Ans.*

Ex. 8. Develop $\frac{1}{\sqrt{c^2-y}}$ into a series. *Ans.*

Ex. 9. Develop $(a^2-x^2)^{-\frac{2}{3}}$ into a series. *Ans.*

The formula of Maclaurin applies in general to all the functions of a single variable that are capable of successive differentiations. But there are cases in which the function or some of its differential coefficients become infinite when $x=o$; in such cases the formula will not apply. The function, $c+ax^{\frac{1}{2}}$ is an example of this kind; for if we represent it by u we have

$$u=c+ax^{\frac{1}{2}}$$

and

$$\frac{du}{dx}=\tfrac{1}{2}ax^{-\frac{1}{2}}=\frac{a}{2x^{\frac{1}{2}}}$$

If in this we make $x=o$ for the value of the coefficient B, we have

$$B=\tfrac{a}{o}=\infty$$

In general no algebraic function of x in which x is not connected with a constant term under the same exponent, can be developed by this theorem; for the differential coefficients will be such as to reduce to zero or infinity in every case, when x is made equal to zero.

TAYLOR'S THEOREM.

(25) *The object of this theorem is to obtain a formula for the development of a function of the sum or difference of two variables.*

The principle on which this theorem is based is the following: The differential coefficient of a function of the sum or difference of two variables, will be the same whether the function is differentiated with respect to one variable alone, or to the other variable alone. Thus the differential coefficient of $(x+y)^n$ will be $n(x+y)^{n-1}$ if we differentiate with respect to either variable alone, the other being considered as constant.

A function of the sum or difference of two variables is one in which both are subject to the same conditions, so that the value of their sum or difference might be expressed by a single variable without otherwise changing the form of the function; and hence we may regard this sum or difference as itself a single variable. Now any rate of change in one of the two component parts (the other being regarded as constant) will produce the same rate in the compound variable (so to speak) as it has itself; thus $x+y$ will increase at the same rate as x if y be constant, and at the same rate as y if x be constant. So that changing from one to the other is merely changing the rate of the single variable that would represent the value of their sum or difference. But such change in the rate will not change the *ratio* which it bears to the corresponding rate of the function (Art. 6); that is, it will not affect the differential coefficient.

(26) To apply this principle let us take any function of the sum of two variables, as $F(x+y)$, which we will represent by u. If it can be developed into a series, the terms of the series may always be arranged according to the powers of y; the coefficients being functions of x and the constants; hence it may be made to take the following form:

$$u=F(x+y)=A+By+Cy^2+Dy^3+Ey^4+\text{ etc.} \qquad (1)$$

in which A, B, C, D, E, etc., are independent of y, but functions of x.

If we differentiate equation (1) regarding y as constant, and divide by dx, we shall have

$$\frac{du}{dx}=\frac{dA}{dx}+\frac{dB}{dx}y+\frac{dC}{dx}y^2+\frac{dD}{dx}y^3+\text{ etc.}$$

If we regard x as constant, and divide by dy, we have

$$\frac{du}{dy}=B+2Cy+3Dy^2+4Ey^3+\text{ etc.}$$

and since $\frac{du}{dx}$ is equal to $\frac{du}{dy}$ the second members of these equations are equal; and since this equality exists for every value of y, and since the coefficients are independent of that value, the corresponding terms containing the same powers of y must be equal each to each; hence

$$\frac{dA}{dx}=B \qquad (2)$$

$$\frac{dB}{dx}=2C \qquad (3)$$

$$\frac{dC}{dx}=3D \qquad (4)$$

$$\frac{dD}{dx}=4E \qquad (5)$$

If now we make $y=o$, then $F(x+y)$ becomes $F(x)$, which we will represent by z. Under this supposition equation (1) will become

$$u \text{ (now become } z) = A$$

Substituting this value of A in equation (2) we have

$$\frac{dz}{dx}=B$$

Substituting this value of B in equation (3) we have

$$\frac{d\frac{dz}{dx}}{dx}=\frac{d^2z}{dx^2}=2C$$

whence

$$C=\frac{d^2z}{2dx^2}$$

similarly we have

$$\frac{d^3z}{2dx^3}=3D \text{ or } D=\frac{d^3z}{2\cdot 3\cdot dx^3}$$

and

$$\frac{d^4z}{2\cdot 3\cdot dx^4}=4E \text{ or } E=\frac{d^4z}{2\cdot 3\cdot 4\cdot dx^4}$$

and so on.

Substituting these values in equation (1) we have

$$u=F(x+y)=z+\frac{dz}{dx}y+\frac{d^2z}{2dx^2}y^2+\frac{d^3z}{2\cdot 3\cdot dx^3}y^3+\text{ etc.}$$

in which the first term is what the function becomes when $y=o$, and all the coefficients of the powers of y are derived from it on the same supposition.

This is the formula of Taylor.

A function of $x-y$ is developed by the same formula by changing y into $-y$, thus:

$$u=F(x-y)=z-\frac{dz}{dx}y+\frac{d^2z}{2dx^2}y^2-\frac{d^3z}{2.3.dx^3}y^3-\text{etc.}$$

EXAMPLES.

Ex. 1. *Let it be* required to develop $(x+y)^n$.

Representing this function by u we have

$$u=(x+y)^n \text{ and } z=x^n$$

then by differentiation

$$\frac{dz}{dx}=nx^{n-1}, \frac{d^2z}{dx^2}=n(n-1)x^{n-2}, \frac{d^3z}{dx^3}=n(n-1)(n-2)x^{n-3}$$

Substituting these values in the formula we have

$u=(x+y)^n=x^n+nx^{n-1}y+\frac{n(n-1)}{2}x^{n-2}y^2+\frac{n(n-1)(n-2)}{2.3.}x^{n-3}y^3+$ etc., the same as by the binomial theorem.

Ex. 2. Develop the function $\sqrt{x+y}$.

Ans. $(x+y)^{\frac{1}{2}}=x^{\frac{1}{2}}+\frac{1}{2}x^{-\frac{1}{2}}y-\frac{1}{2\cdot4}x^{-\frac{3}{2}}y^2+\frac{1\cdot3}{2\cdot4\cdot6}x^{-\frac{5}{2}}y^3-$ etc.

Ex. 3. Develop $\sqrt[3]{x+y}$.

Ans. $x^{\frac{1}{3}}+\frac{1}{3}x^{-\frac{2}{3}}y-\frac{2}{3\cdot6}x^{-\frac{5}{3}}y^2+\frac{2\cdot5}{3\cdot6\cdot9}x^{-\frac{8}{3}}y^3-$ etc.

Ex. 4. Develop the function $(x-y)^n$. *Ans.*

Ex. 5. Develop the function $(x-y)^{\frac{3}{2}}$. *Ans.*

Ex. 6. Develop the function $\frac{1}{x-y}$. *Ans.*

Ex. 7. Develop the function of $a(x-y)^{-\frac{2}{3}}$. *Ans.*

(**27**) Although a function of the sum or difference of two variables can generally be developed by this formula, yet there are cases in which the coefficients (which are functions of one of the variables) may, by giving certain values to the variable they contain, become infinite. In such cases the formula cannot be applied; for in general such values for that variable, would not reduce the function itself to infinity, although it would have that effect on its development. Thus in the function

$$u=a+\ (b-x+y)^{\frac{1}{2}}$$

we have

$$z=a+(b-x)^{\frac{1}{2}},\ \frac{dz}{dx}=\frac{1}{2(b-x)^{\frac{1}{2}}},\ \frac{d^2z}{dx^2}=-\frac{1}{4(b-x)^{\frac{3}{2}}}$$

and so on.

If now we make $x=b$, all the coefficients will become infinite, and we should have

$$u=a+(b-x+y)^{\frac{1}{2}}=a+\infty$$

by the formula instead of having as we ought

$$u=a+y^{\frac{1}{2}};$$

which cannot be, for the value of y is not dependent on that of x, and hence u is not necessarily infinite when $x=b$; but for all other values of x the formula will give the true development of the function.

And herein is the difference between the formulas of Taylor and Maclaurin; when that of the former fails it is for only one value of the variable; while that of the latter, when it fails at all, fails for every value of it.

NOTE.—In fact, the theorems of both Taylor and Maclaurin are founded on the principle illustrated in Art. 21. The real object of both is to find from the rate of change of a function what will be its new state arising from a given change in the value of the variable.

The general method of doing this is to find the successive differentials of the function in its first state, and then to multiply the successive differential coefficients by the successive powers (properly divided) of the actual change in the variable; this will give the actual successive partial changes in the function which together make up the entire change, and thus develop the function in the new state.

For this purpose the variable must have two points of value; one where the function is to be differentiated, and the other, the new value produced by the change; and to this end the variable, in algebraic functions, is made to consist of two parts, either by making it a binomial or something that may be reduced to that form.

In Maclaurin's theorem the variable consists of a constant and a variable combined together, so that their united value is a variable one, and the constant part is simply one point in that variable value. This is the point at which the differentiation of the function is made; but as a constant cannot be differentiated the variable is attached to it long enough for that purpose and then made zero. In Taylor's theorem the variable is the sum or difference of two others, and the point of differentiation is when the variable has reached the value of one of its variable parts. This being a variable, the function can be differentiated directly, and the other variable may be made zero *before* the operation. Hence the theorems of Maclaurin and Taylor are alike in this; both have a compound variable having two points of value, both are differentiated at the same point, and the successive differential coefficients, which are precisely alike in form and value, are multiplied by the successive powers of the change in value. The only difference is that in one the differentiation at the required point is made *indirectly*, and the variable change made zero afterwards; while in the other the differential is made *directly*, the variable change being made zero beforehand. Hence a function of a binomial variable may be expanded by either method. By Taylor's, considering both terms variable and reducing one to zero *before* differentiation; or, by Maclaurin's, by considering one term as constant and reducing the other to zero *after* differentiation. Thus in the case of the function $(x+y)^n$, the differential coefficient will be precisely the same if we reduce y to zero and differentiate x by Taylor's method, or consider x as constant and reduce y to zero after differentiation, by Maclaurin's method.

In order that a binomial may represent a single variable, both terms must be subject to the same conditions, so that each term may be considered as a part of the same compound variable; and the failing cases in Maclaurin's and Taylor's methods are simply those in which the binomial variable becomes a monomial, by giving the variable a certain value. Thus the case cited in Art. 24, $c+ax^{\frac{1}{2}}$, is not a true binomial variable, since the terms are not subjected to the same conditions. If we make it $(c+ax)^{\frac{1}{2}}$ we have a true binomial variable, and the differential coefficient $\frac{1}{2}(c+x)^{-\frac{1}{2}}$ will *not* reduce to infinity when $x=o$. Similarly the case cited in Art. 27, namely $u=a+(b-x+y)^{\frac{1}{2}}$ is one in which when $x=b$, the variable in the function reduces to y, and the function itself to $a+y^{\frac{1}{2}}$, which does not contain a binomial variable of the required form.

The same principle will apply to transcendental functions; which, in order to be developed, must have two points of value in the compound variable—one for the differential and the other for the development. Thus $a^x k$ may be expanded by Maclaurin's theorem, since it has two points of value, one at k, the point of differentiation where $x=o$, and the other the full value produced by x.

SECTION IV.

MAXIMA AND MINIMA.

(28) We have seen that when a variable changes its value at a uniform rate, the value of its function will in general change at a rate that is not uniform. It may increase at a diminishing rate, until at a certain point it ceases to increase and begins to diminish, in which case the turning point is the one of greatest value, and is called a maximum. Or it may decrease to a certain point and, having attained its minimum, begin to increase. The problem is to find whether there *is* a maximum or minimum value for a function, while its variable is uniformly increasing, and if there is, to find the corresponding value of the variable and of its function.

(29) While a positive function is increasing as the variable increases, its rate of change or differential will be positive; and negative when it decreases (Art. 3). Hence when a function is passing through a maximum or minimum value, the sign of the differential coefficient must change from minus to plus or from plus to minus — the former in case of a minimum, and the latter in case of a maximum.

But such a change can take place only while the differential coefficient is passing through zero or infinity. Our first inquiry then is whether there is any finite value of the variable that will reduce the first differential coefficient to either of these values. For this purpose we solve the equation formed by placing the first differential coefficient equal to zero, and thus find the corresponding value of the variable. Here we have one of three results.

First. There may be no real value for the variable. In this case there is neither maximum nor minimum.

Second. There may be a real finite value for the variable that will reduce the differential coefficient to zero. In this case there will *probably* be a maximum or minimum.

Third. There may be no finite value of the variable that will reduce the differential coefficient to *zero*, but at the same time there may be one that will reduce it to *infinity*. In this case we form the equation by placing the differential coefficient equal to infinity, and the root that satisfies the equation will indicate a *probable* maximum or minimum.

In order to determine in the two latter cases whether there *is* a maximum or minimum value of the function, and if so, which of the two it is, we may substitute in the function, in place of the variable, a quantity a little less, and one a little greater than that derived from the equation. If the result in both cases is less than when the true value is substituted there is a maximum; if greater, there is a minimum value of the function for the true value of the variable.

We may also determine the same thing by substituting these approximate values in the differential coefficient, which the true value reduces to zero. If they cause the result to change the sign from plus to minus by substituting first the less, and then the greater quantity, there is a maximum, for the function is passing from an increasing to a decreasing state. If the change is from minus to plus, there is a minimum, for the function is passing from a decreasing to an increasing state at that point.

EXAMPLE.

Find the value of x which will render u a maximum or minimum in the equation,

$$u=x^3-9x^2+24x-7$$

Differentiating and placing the differential coefficient equal to zero we have

$$\frac{du}{dx}=3x^2-18x+24=3(x^2-6x+8)=0$$

from which we find

$$x=4 \text{ and } x=2$$

If we substitute in the function and in the differential coefficient 1, 2, 3, 4, 5, etc., successively, we shall have for

$$x=1 \quad . \; . \; . \quad u=9 \quad . \; . \; . \quad \frac{du}{dx}=3$$
$$x=2 \quad . \; . \; . \quad u=13 \quad . \; . \; . \quad \frac{du}{dx}=0$$
$$x=3 \quad . \; . \; . \quad u=11 \quad . \; . \; . \quad \frac{du}{dx}=-1$$
$$x=4 \quad . \; . \; . \quad u=9 \quad . \; . \; . \quad \frac{du}{dx}=0$$
$$x=5 \quad . \; . \; . \quad u=13 \quad . \; . \; . \quad \frac{du}{dx}=3$$
$$x=6 \quad . \; . \; . \quad u=29 \quad . \; . \; . \quad \frac{du}{dx}=24$$

Indicating that for $x=2$ the value of the function is a maximum, the differential coefficient passing from plus to minus; and for $x=4$ the value of the function is a minimum—the differential coefficient passing from minus to plus.

(30) It must be understood that by maximum and minimum is not meant the absolutely greatest or least value of the function, but the *turning point*, from an increase to a decrease, or *vice versa*. Hence there may be as many maxima or minima of the function as there are values of the variable that will reduce the first differential coefficient to zero or infinity.

It is also to be understood that in the discussion of *this subject*, when a function is stated to be an *increasing* one, it is meant that it is either positive and becoming greater, or negative and becoming less. If it is said to be *decreasing*, it is either positive and becoming less or negative and becoming greater. Thus if we take the function

$$u=x^2-25$$

and make x, successively, equal to

$$1 . 2 . 3 . 4 . 5 . 6 . 7 .$$

the successive values of the function will be

$$-24, -21, -16, -9, 0 +11, +24$$

and it is said to be increasing throughout the whole change, although at first its numerical negative values decrease. This is also indicated by the sign of the differential coefficient which is positive as long as x is positive. So that the maximum value of the function may be the greatest positive or the least negative value, and the minimum the least positive or the greatest negative value.

(31) There is another method of ascertaining whether the first differential coefficient changes its sign on passing through zero or infinity, for this is the unfailing test of a maximum or minimum. Having found that value of the variable which reduces the first differential coefficient to zero, substitute that value in the second differential coefficient, if it contain the variable; then

First. If it reduces the second differential coefficient to a *negative* quantity, it indicates that when the first is at zero it must be a decreasing function, which can be at that point only by its passing *from a positive to a negative state*, and hence the function itself must be passing from a state of increase to a state of decrease, and hence is at a maximum.

Second. If it reduces the second differential coefficient to a *positive* quantity, it indicates that the first when at zero is an increasing function, and must, therefore, be passing *from a negative to a positive state*, hence the function is passing from a decreasing to an increasing state. and is, therefore, at a minimum.

Third. If it reduces the second differential coefficient to zero, we may resort to the third; and if the same value of the variable reduces that to a real finite quantity, either positive or negative, it shows that the second, at zero, is changing its sign, and, therefore, the first is changing from an increasing to a diminishing function, or *vice versa*, and, therefore, does *not at the zero point change its sign.* Hence there is neither maximum nor minimum in the value of the function.

Fourth. If it reduces the *third* differential coefficient to zero we may resort to the *fourth.* If it reduces this to a real finite value, it indicates that at zero the third changes its sign, for it can increase on both sides of zero only by passing from negative to positive, or diminish on both sides by passing from positive to negative. This will show that the second coefficient does *not* change its sign, for if it increases before it becomes zero and decreases afterwards, or *vice versa*, it can approach the zero point only until it *touches it*, and then must recede without changing its sign. This proves that the first coefficient *does* change its sign, for since the second does not change the first must be passing *through* from one side to the other. There will, therefore, be a maximum or minimum—the first if the fourth differential coefficient has a negative value. and the second if it is positive.

We may continue thus and show that if the first differential coefficient that is reduced to a real value, by substituting that value of the variable that reduces the first to zero, is of an *even order* and *positive*, there will be a *minimum;* if it is *negative*, there will be a *maximum;* and if it is of an *odd order* there will be *neither maximum nor minimum.*

Fifth. If any value of the variable reduces the first differential coefficient to infinity, it w ll probably reduce all the succeeding ones to infinity, also. It will, therefore, be best in such a case to substitute values for the variable a little less and then a little greater than the one found. If the value of the first differential coefficient changes from plus to minus there is a maximum, and the second will be plus on both sides of infinity; for the first must be an *increasing* positive function in order to become positively infinite, and if negative on the other side, must be a decreasing function, for it *cannot* be an *increasing* negative function on leaving infinity. Hence (Art. 3) the second must be positive in both cases.

Sixth. If the first differential coefficient in the last case changes from minus to plus there will be a minimum, and the second coefficient will be minus on both sides of infinity. Thus we see that when any value of the variable reduces the first differential coefficient to *zero*, and is substituted in the second, a *minus* result indicates a *maximum* in the function, and a *plus* result a *minimum*. When any value reduces the first coefficient to *infinity*, a *plus* sign for the second indicates a *maximum*, and a *minus* sign a *minimum*.

EXAMPLES.

Ex. 1. In order to illustrate the first case in this article we take the function

$$u=16x-x^2 \qquad (1)$$

from which we obtain by differentiation

$$\frac{du}{dx}=16-2x \qquad (2)$$

$$\frac{d^2u}{dx^2}=-2 \qquad (3)$$

We find that $x=8$ will reduce the first differential coefficient to zero, while the sign of the second is minus. Hence at $x=8$ the first must be a decreasing function, and, therefore, passing through zero from plus to minus. the function will, therefore, be an increasing one to that point and then a diminishing one; hence a maximum.

If we substitute in the function and the first differential coefficient for x values a little less and a little greater than 8, we have for

$$x=7 \;..\; u=63 \;.\; \frac{du}{dx}=2$$
$$x=8 \;...\; u=64 \;...\; \frac{du}{dx}=0$$
$$x=9 \;....\; u=63 \;....\; \frac{du}{dx}=-2$$

If we represent the values of the function by the ordinates of the curve A B C (Fig. 3), the curve itself will correspond to the range or locus of values of the function, while the variable increases uniformly in passing from 7 to. 9. From A to B the function increases, but at a decreasing rate, and hence the first differential coefficient is positive but decreasing until it reaches zero at B. The function then decreases at an increasing rate, and hence the first differential coefficient must be negative and increasing.

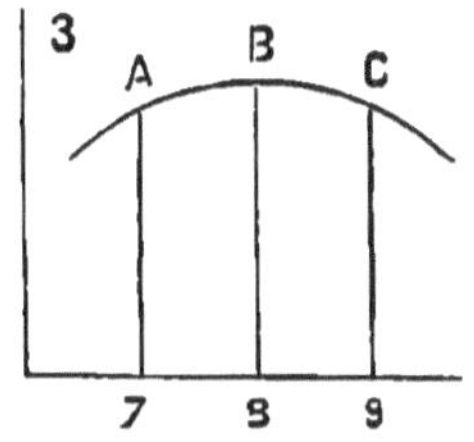

But when this coefficient (or any variable) is positive and decreasing, or negative and increasing, *its rate of change*, *i.e.*, the second differential of the function, must (Art. 3) be negative throughout, which corresponds with the result found in equation (3).

Ex. 2. To illustrate the second case we take

$$u=x^2-16x+70 \tag{1}$$

from which

$$\frac{du}{dx}=2x-16 \tag{2}$$

$$\frac{d^2u}{dx^2}=2 \tag{3}$$

We infer from equation (3) that $\frac{du}{dx}$ is an increasing function for all values of x, and hence it is so when $x=8$, which reduces $\frac{du}{dx}$ to zero. From which we infer that $\frac{du}{dx}$ is passing from a negative to a positive state, and the function itself from a decrease to an increase. Hence a minimum.

If we substitute 7, 8 and 9 successively for x in equations (1) and (2), we have for

$$x=7 \quad u=7 \quad \frac{du}{dx}=-2$$

$$x=8 \quad u=6 \quad \frac{du}{dx}=0$$

$$x=9 \quad u=7 \quad \frac{du}{dx}=2$$

which corresponds with our deductions.

If we let the ordinates of the curve A B C (Fig. 4) represent the values of u, we see that from A to B the value of u diminishes, as is shown by the sign of $\frac{du}{dx}$, and at a diminishing rate as is shown by the positive sign of $\frac{d^2u}{dx^2}$ (Art. 3). From B to C u increases, as is shown by the positive sign of $\frac{du}{dx}$, and at an increasing rate, as is shown by the sign of $\frac{d^2u}{dx^2}$, which is still plus. Hence the shape of the curve.

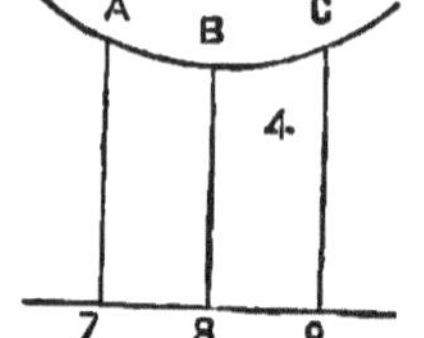

Ex. 3. To illustrate the third case we make

$$u=9+2(x-3)^3 \tag{1}$$

whence

$$\frac{du}{dx}=6(x-3)^2 \tag{2}$$

$$\frac{d^2u}{dx^2}=12(x-3) \tag{3}$$

Here we find $x=3$ reduces $\frac{du}{dx}$ to zero, and hence if there is a maximum or minimum it will be for that value of x.

But it also reduces $\frac{d^2u}{dx^2}$ to zero, hence we resort to the value of $\frac{d^3u}{dx^3}$, which we find to be 12. We infer from this that when $\frac{d^2u}{dx^2}=0$ it is passing from negative to positive, hence $\frac{du}{dx}$ is passing from a decreasing to an increasing function at the zero point, and, therefore, does *not* pass through it. It is, therefore, all the time positive, and the function is at all times an increasing one, so that there is neither maximum nor minimum. We may, also, learn the same thing from inspection, for since the value of $\frac{du}{dx}$ is a square it must always be positive.

If we substitute in the given function and in the values of $\frac{du}{dx}$ and $\frac{d^2u}{dx^2}$ the numbers 2, 3 and 4 successively for x, we have for

$$x=2 \quad u=7 \quad \frac{du}{dx}=6 \quad \frac{d^2u}{dx^2}=-12$$

$$x=3 \quad u=9 \quad \frac{du}{dx}=0 \quad \frac{d^2u}{dx^2}=0$$

$$x=4 \quad u=11 \quad \frac{du}{dx}=6 \quad \frac{d^2u}{dx^2}=12$$

If we let the ordinates of the curve A B C (Fig. 5) represent the values of u, we see that from 2 to 3 the function increases, as is shown by the positive value of $\frac{du}{dx}$, but at a decreasing rate, as is indicated by the negative sign of $\frac{d^2u}{dx^2}$ (Art. 3), between those points or while x is less than 3. From 3 to 4 the function is still increasing, as is

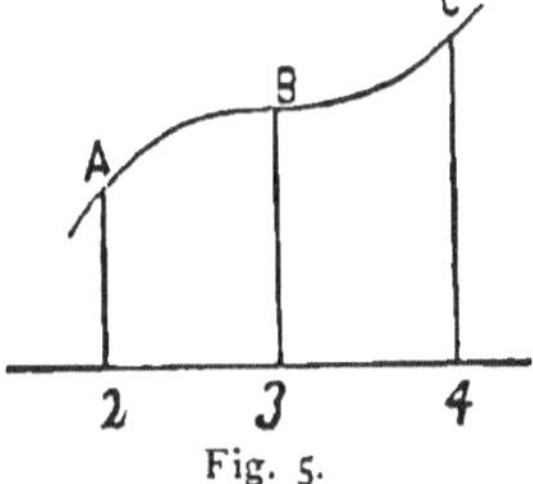

Fig. 5.

shown by the positive sign of $\frac{du}{dx}$, and at an increasing rate, as is shown by the positive sign of $\frac{d^2u}{dx^2}$ when the value of x is greater than 3. Hence at B where the value of $\frac{du}{dx}$ is zero, the function having increased at a decreasing rate up to that point, ceases for an inappreciable moment, and begins again to increase at an increasing rate.

Ex. 4. To illustrate the fourth case take

$$u=5+(x-7)^4 \qquad (1)$$

whence

$$\frac{du}{dx}=4(x-7)^3 \qquad (2)$$

$$\frac{d^2u}{dx^2}=12(x-7)^2 \qquad (3)$$

$$\frac{d^3u}{dx^3}=24(x-7) \qquad (4)$$

$$\frac{d^4u}{dx^4}=24 \qquad (5)$$

Here we see that the fourth differential coefficient is the first that has a real finite value for $x=7$, which reduces all the preceding ones to zero. Hence, according to our rule, there should be a minimum value for the function at that point. In fact, the sign of this coefficient shows that the third changes at zero from minus to plus; and this shows that the second does *not* change its sign at zero, but after being a decreasing function to that point, becomes an increasing one, and is, therefore, positive both before and after. And this again shows that the first *does* change its sign at zero, since it is an increasing function on both sides of zero, which can be only by passing from a negative to a positive value. Hence the function will decrease until $x=7$, after which it will increase, showing a minimum at that point.

If we substitute in the given function and in the differential coefficients the numbers 5, 6 7, 8, 9 successively for x we shall have for

$$x=5 \quad u=21 \quad \frac{du}{dx}=-32 \quad \frac{d^2u}{dx^2}=48 \quad \frac{d^3u}{dx^3}=-48$$
$$x=6 \quad u=6 \quad \text{"}=-4 \quad \text{"}=12 \quad \text{"}=-24$$
$$x=7 \quad u=5 \quad \text{"}=0 \quad \text{"}=0 \quad \text{"}=0$$
$$x=8 \quad u=6 \quad \text{"}=4 \quad \text{"}=12 \quad \text{"}=24$$
$$x=9 \quad u=21 \quad \text{"}=32 \quad \text{"}=48 \quad \text{"}=48$$

which illustrates the conclusions we have drawn.

The general proposition enunciated in the fourth case may be demonstrated analytically as follows. Let us suppose

$$u=F(x)$$

and let the variable x be first increased and then diminished by another variable h; and let these new states of the function be represented by u' and u'', then we have

$$u'=F(x+h)$$
$$u''=F(x-h)$$

Developing these by Taylor's theorem we have, after sub tracting the original function u,

$$u'-u=\frac{du}{dx}h+\frac{d^2u}{dx^2}\cdot\frac{h^2}{2}+\frac{d^3u}{dx^3}\cdot\frac{h^3}{2.3}+\text{ etc.}$$
$$u''-u=-\frac{du}{dx}h+\frac{d^2u}{dx^2}\cdot\frac{h^2}{2}-\frac{d^3u}{dx^3}\cdot\frac{h^3}{2.3}+\text{ etc.}$$

Since the powers of h increase in each successive term of this development, we may reduce the value of h to such an extent that the value of any one term shall be greater than that of all the succeeding terms added together. Such in fact will be the case if h is less than one-half in the series h, h^2, h^3, etc.

Let us suppose h to be so reduced, then if u is a maximum, it must be greater than u' or u'', and the second members of both these equations must be negative; if it is a minimum it must be less than u' or u'', and in this case the second members of both equations must be positive.

Hence, in case of a maximum or minimum, the second members of both equations must have the same sign, and the first term of each (which controls the value of all the rest) having contrary signs must reduce to zero; that is, $\frac{du}{dx}$ must be zero, since h is not. This then is a necessary condition to a maximum or minimum. If there is a real value for the second term in each equation, its sign (or that of $\frac{d^2u}{dx^2}$ since h^2 is positive), will now control that of the whole second member, and will determine whether u is a maximum or minimum. If the second term (or $\frac{d^2u}{dx^2}$) become zero, there can be no maximum or minimum unless the third term (or $\frac{d^3u}{dx^3}$) which is now the controlling term, is also zero, since it has contrary signs in the two equations. We see then that the conditions of a maximum or minimum are: *first*, that the first differential coefficient should become zero; and, *second*, that the first succeeding differential coefficient that has a real value should be one of an *even* order, since the even terms have the same sign in both equations. If that is negative, the whole of the second member of the equation is negative, and there is a maximum; if it is positive, there is a minimum. Which agrees with the rule already found

Ex. 5. To illustrate the fifth case we take

$$u=10-(x-3)^{\frac{2}{3}} \tag{1}$$

whence

$$\frac{du}{dx}=\frac{-2}{3(x-3)^{\frac{1}{3}}} \tag{2}$$

$$\frac{d^2u}{dx^2}=\frac{2}{9(x-3)^{\frac{4}{3}}} \tag{3}$$

Here we find that $x=3$ will reduce $\frac{du}{dx}$ to infinity. Referring to the value of $\frac{d^2 u}{dx^2}$ we find that it reduces that also to infinity, but we see by inspection that any other value for x, whether greater or less than 3, will make that of $\frac{d^2 u}{dx^2}$ positive. We see also that $\frac{du}{dx}$ will be positive when x is less than 3, and negative when it is greater. From all this we infer that the function is an increasing one before $x=3$, and a decreasing one afterwards. Hence there is a maximum at that point.

If we substitute for x, in equations (1), (2) and (3), the numbers 2 . 3 . 4 successively, we have for

$$x=2 \quad u=\ 9 \quad \frac{du}{dx}=\tfrac{2}{3} \quad \frac{d^2 u}{dx^2}=\tfrac{2}{9}$$

$$x=3 \quad u=10 \quad \frac{du}{dx}=\infty \quad \frac{d^2 u}{dx^2}=\infty$$

$$x=4 \quad u=\ 9 \quad \frac{du}{dx}=-\tfrac{2}{3} \quad \frac{d^2 u}{dx^2}=\tfrac{2}{9}$$

If we let the ordinates of the curve A B C represent the successive values of u (Fig 6.) we see that from 2 to 3 the function increases, as is shown by the positive value of $\frac{du}{dx}$ and at an increasing rate as is shown by the positive value of $\frac{d^2 u}{dx^2}$. From 3 to 4 the negative sign of $\frac{du}{dx}$ indicates a decrease of the function, while the positive sign of $\frac{d^2 u}{dx^2}$ (which has not changed) shows that this decrease is at a decreasing rate. Hence the form of the curve.

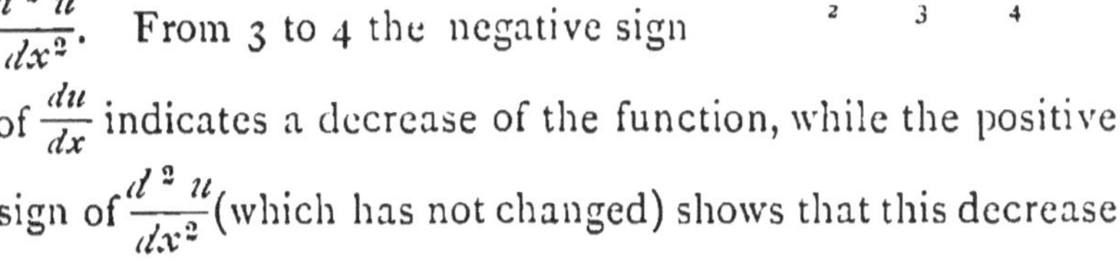

Ex. 6. To illustrate the sixth case we take

$$u=(3x-9)^{\frac{2}{3}} \tag{1}$$

whence

$$\frac{du}{dx}=\frac{2}{(3x-9)^{\frac{1}{3}}} \tag{2}$$

$$\frac{d^2u}{dx^2}=-\frac{2}{(3x-9)^{\frac{4}{3}}} \tag{3}$$

In this case, as before, we find that $x=3$ reduces $\frac{du}{dx}$ and $\frac{d^2u}{dx^2}$ to infinity. We see also by inspection that $\frac{du}{dx}$ changes from minus to plus as x passes from less than 3 to greater, while $\frac{d^2u}{dx^2}$ is negative on both sides of infinity. Hence we infer that u is a decreasing function until $x=3$, and an increasing one afterwards. Hence a minimum at that point.

If we substitute for x in equations (1), (2), (3), the numbers 2 . 3 . 4 successively we shall have for

$$x=2 \quad u=\sqrt[3]{9} \quad \frac{du}{dx}=-\frac{2}{\sqrt[3]{3}} \quad \frac{d^2u}{dx^2}=-\frac{2}{\sqrt[3]{81}}$$

$$x=3 \quad u=0 \quad \frac{du}{dx}=\infty \quad \frac{d^2u}{dx^2}=\infty$$

$$x=4 \quad u=\sqrt[3]{9} \quad \frac{du}{dx}=\frac{2}{\sqrt[3]{3}} \quad \frac{d^2u}{dx^2}=-\frac{2}{\sqrt[3]{81}}$$

If we let the ordinates of the curve A B C (Fig. 7) represent the successive values of u we see that from 2 to 3 the function decreases, as is shown by the negative value of $\frac{du}{dx}$, and at an increasing rate, as is shown by the negative value of $\frac{d^2u}{dx^2}$; from 3 to 4 the positive value of $\frac{du}{dx}$ indicates an increasing function, while the

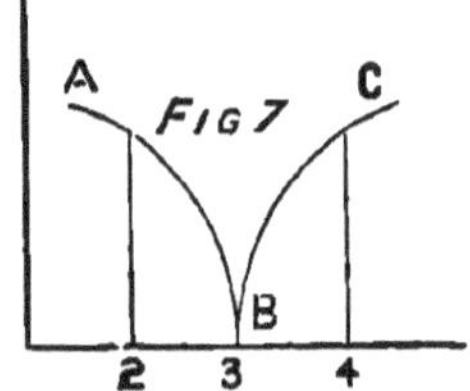

negative value of $\frac{d^2u}{dx^2}$ shows that increase to be at a diminishing rate. Hence the form of the curve.

(32) We have seen (Art. 30) that there may be as many maxima and minima of a function as there are roots for the equation formed by making the value of $\frac{du}{dx}=o$. To illustrate a case of this kind we take

Ex. 7.

$$u=x^4-20x^3+132x^2-320x+286 \qquad (1)$$

whence

$$\frac{du}{dx}=4x^3-60x^2+264x-320 \qquad (2)$$

$$\frac{d^2u}{dx^2}=12x^2-120x+264 \qquad (3)$$

Placing the second member of equation (2) equal to zero, we find for x three values as follows:

$$x=2$$

$$x=5$$

$$x=8$$

Substituting these values in equation (3) we have for

$$x=2 \quad \frac{d^2u}{dx^2}=72$$

$$x=5 \quad \frac{d^2u}{dx^2}=-36$$

$$x=8 \quad \frac{d^2u}{dx^2}=72$$

from which we infer that for $x=2$ and $x=8$ there is a minimum, and for $x=5$ there is a maximum. This will be seen by substituting for x in equations (1), (2), (3), successive values, as follows:

x	u	$\frac{du}{dx}$	$\frac{d^2u}{dx^2}$
$x=0$	$u=286$	$\frac{du}{dx}=-320$	$\frac{d^2u}{dx^2}=264$
$x=1$	$u=79$	" $=-112$	" $=156$
$x=2$	$u=30$	" $=0$	" $=72$
$x=3$	$u=55$	" $=40$	" $=12$
$x=4$	$u=94$	" $=32$	" $=-24$
$x=5$	$u=111$	" $=0$	" $=-36$
$x=6$	$u=94$	" $=-32$	" $=-24$
$x=7$	$u=55$	" $=-40$	" $=12$
$x=8$	$u=30$	" $=0$	" $=72$
$x=9$	$u=79$	" $=112$	" $=156$
$x=10$	$u=286$	" $=320$	" $=264$

If we let the ordinates of the curve A B C (Fig. 8) rep-

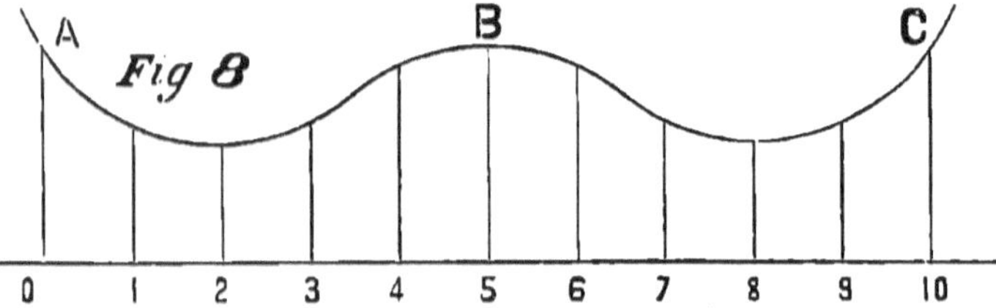

resent the successive values of u corresponding to numbers substituted for x at the foot of each, we see that the function decreases at a diminishing rate until $x=2$, when it ceases to decrease and begins to increase at an increasing rate, as is shown by the change of sign in $\frac{du}{dx}$ and the continued positive sign of $\frac{d^2u}{dx^2}$. But at $x=4$, although still increasing, as is shown by the positive sign of $\frac{du}{dx}$, it is at a diminishing rate, for $\frac{d^2u}{dx^2}$ is now negative, and thus continues until at 5, $\frac{du}{dx}$ becomes zero, the function ceases to increase and begins to diminish at an increasing rate, as is shown by the negative signs of $\frac{du}{dx}$ and $\frac{d^2u}{dx^2}$ at $x=6$. But at $x=7$ we have $\frac{d^2u}{dx^2}$ positive, showing that the rate of diminution of the function has ceased to increase and begins to diminish, until at $x=8$ it has become zero, when the changes at $x=2$ are repeated. We notice that between $x=3$ and $x=4$ the sign of $\frac{d^2u}{dx^2}$ changes from plus to minus, showing that between those two points the rate of $\frac{du}{dx}$ has changed from an increase to a decrease, that is, the function has changed from increasing at an increasing rate to increasing at a diminishing rate. The exact point where this change takes place is where the value of $\frac{d^2u}{dx^2}=0$. This will give

$$x=5-\sqrt{3} \text{ and } x=5+\sqrt{3}$$

which last value corresponds to a point between $x=6$ and $x=7$, where the same change is repeated, only in a contrary direction.

From all these cases we deduce the following rule for ascertaining the values of the variable that will produce a maximum or minimum value for the function, if there be any.

Place the first differential coefficient equal to zero; and substitute each of the roots of this equation for the variable in the second differential coefficient. Each one that reduces it to a real negative quantity will produce a maximum value for the function; while a similar positive result will indicate a minimum. Should any real root thus found reduce the second differential coefficient to zero, substitute it in the third, fourth, etc., successively, until a real finite value is found for some one of the coefficients. If the first thus found be of an even order and positive, there will be a minimum; if negative there will be a maximum. If the first that is reduced to a real finite quantity is of an odd order, whether positive or negative, there will be neither maximum nor minimum.

The first differential coefficient may also be placed equal to infinity, and if there be any real finite roots, they may be treated in the same manner as those obtained by placing it equal to zero. In this case, however, a positive sign of the second differential coefficient indicates a maximum and a negative sign a minimum.

If a given function contains two variables there must be an equation, and one of the variables must be considered as dependent on the other. The problem will be to find the maximum or minimum value of the dependent variable; for which purpose it must be considered as an implicit function of the other, and the differential coefficients will be found as in other cases.

NOTE.—It may be objected that, herein, the subject of maxima and minima has been treated in too prolix a manner, and the reasoning has been unnecessarily repeated. I reply, it is of the highest importance that the student should have not only a clear and correct, but a *familiar*, conception of the *laws* which govern the relations of the different orders of rates or differentials, because these are among the fundamental ideas of the calculus, and essential to a complete comprehension of the subject. Now unless these ideas are presented sufficiently often to render them familiar; if the student on every new occasion is obliged to draw afresh upon his powers of imagination, and go through the mental labor of forming his conceptions anew, the study will prove not only more difficult, but far less attractive. He will be like a traveler in the dark, who, instead of carrying a constantly shining lamp to guide his footsteps, must light his candle anew for every fresh obstacle. Hence the importance of a full and elaborate explanation, even at the expense of some, otherwise unnecessary, repetition.

EXAMPLES

Find the value of x for the maximum or minimum value of u in the following equations:

Ex. 8. $u=x^3+18x^2+105x$. *Ans.*

Ex. 9. $u=a-bx+x^2$. *Ans*

Ex. 10. $u=a^4+b^3x-cx^2$. *Ans.*

Ex. 11. $u=3a^2x^3-b^4x+c^2$. *Ans.*

Ex. 12. $u=a^2+bx^2-cx^3$. *Ans.*

APPLICATION TO PRACTICAL PROBLEMS.

(33) In order to apply the rules for determining maxima and minima of functions to the solution of practical problems, it is necessary to obtain an algebraic expression of the function, whose maximum or minimum is to be determined, in such terms that it shall contain but one variable. No specific rules can be given for this purpose, but care must be taken to express the function in terms of a variable that shall have a range of values *beyond* that which may be required to produce a maximum or minimum, for if it does not, although there may be a kind of maximum or minimum, it will not be one in the meaning of the term as used in the calculus, as there will be no *turning point* in the value of the function, nor any change of sign in the value of the first differential coefficient. A few examples will indicate the nature of the process more clearly.

Ex. 1. Divide the quantity a into two such parts that their product shall be a maximum.

Let x be one of these parts, then the other will be $a-x$, and the function will be

$$x(a-x)=ax-x^2$$

which is to be a maximum. Representing it by u we have

$$\frac{du}{dx}=a-2x \qquad (1)$$

$$\frac{d^2u}{dx^2}=-2 \qquad (2)$$

Placing the value of $\frac{du}{dx}$ equal to zero we have

$$x=\frac{a}{2}$$

which the negative sign of $\frac{d^2u}{dx^2}$ shows to be a maximum.

Hence *when a quantity is divided into two parts their product is a maximum when they are equal.*

Ex. 2. To find the greatest cylinder that can be inscribed in a given right cone.

Let the height SC (Fig. 9) of the cone be represented by a, and the radius of the base AC by b, and let x represent the distance SD from the vertex of the cone to the upper base of the cylinder. From the triangles SAC and SED we have SC : AC :: SD : ED or $a : b :: x$: ED, hence

$$ED = \frac{bx}{a}$$

But the area of the upper base of the cylinder is

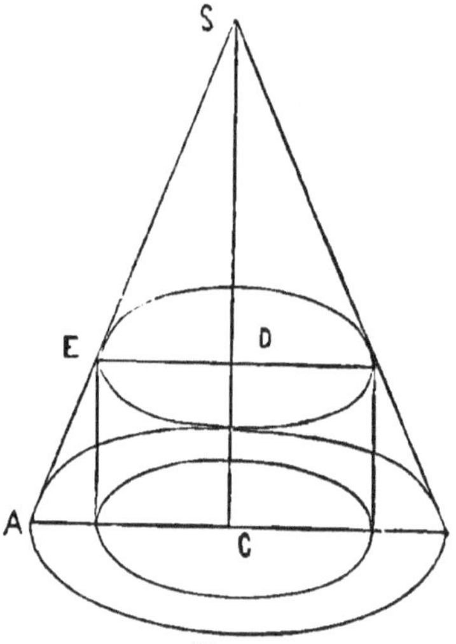

Fig. 9.

$$\pi ED^2 = \pi \frac{b^2 x^2}{a^2}$$

Multiplying this by $DC = a - x$, the height of the cylinder, we have the volume or capacity, which we will represent by V, and hence

$$V = \frac{\pi b^2}{a^2} x^2 (a - x)$$

Now any value of x that will render $x^2(a-x)$ a maximum will render any multiple of it also a maximum, hence $\frac{\pi b^2}{a^2}$ being a constant factor may be disregarded in the operation. Differentiating twice and representing $x^2(a-x)$ by u, we have

$$\frac{du}{dx} = 2ax - 3x^2$$

$$\frac{d^2 u}{dx^2} = 2a - 6x$$

Making the value of $\frac{du}{dx}$ equal to zero we have

$$x = 0 \text{ and } x = \frac{2a}{3}$$

The first cannot be a maximum since it reduces the value of $\frac{d^2u}{dx^2}$ to $2a$, which being positive indicates a minimum. In fact, when $x=o$ the cylinder is reduced to the axis of the cone, and vanishes with x. The other value $x=\frac{2}{3}a$ will solve the problem, since it reduces the value of $\frac{d^2u}{dx^2}$ to $-2a$ a negative quantity which indicates a maximum value for the function. Hence

The maximum cylinder that can be inscribed in a right cone is one in which the height of the cylinder is one-third of the height of the cone. The radius of the base will also be equal to two-thirds that of the base of the cone. The volume of the cylinder will be to that of the cone in the ratio of their bases, or as 4 is to 9.

Ex. 3. Required to determine the dimensions of a cylindrical vase, that will contain a given quantity of water with the least amount of surface in contact with it.

Let v represent the given volume of water, x the radius of the base of the cylindrical vase, and y its altitude. Then we shall have

$$v=\pi x^2 y$$

from which

$$y=\frac{v}{\pi x^2}$$

Now the convex surface of the cylinder is equal to $2\pi xy$, and substituting the value of y we shall have the convex surface equal to

$$\frac{2\pi xv}{\pi x^2}=\frac{2v}{x}$$

If to this we add the surface of the base $=\pi x^2$ we have the whole surface in contact with the water. Calling this surface S we have

$$S=\frac{2v}{x}+\pi x^2 \tag{1}$$

$$\frac{dS}{dx}=-\frac{2v}{x^2}+2\pi x \tag{2}$$

$$\frac{d^2S}{dx^2}=\frac{4v}{x^3}+2\pi \tag{3}$$

Placing the value of $\frac{dS}{dx}$ equal to zero we have

$$2\pi x^3=2v$$

whence

$$x=\sqrt[3]{\frac{v}{\pi}}$$

This value answers to a minimum since it renders the value of $\frac{d^2S}{dx^2}$ positive. If we substitute this value of x in the expression we found for the value of y, namely

$$y=\frac{v}{\pi x^2}$$

we have

$$y=\frac{v}{\pi\sqrt[3]{\frac{v^2}{\pi^2}}}=\sqrt[3]{\frac{v}{\pi}}$$

hence *the minimum surface will be in contact with the water when the height of the cylinder is equal to the radius of the base.*

Ex. 4. It is required to inscribe in a sphere a cone which shall have the greatest convex surface.

Suppose the semi-circumference AMB (Fig. 10) to revolve about the axis AB, it will describe the surface of a sphere, and the chord AM will describe the convex surface of a right cone inscribed in the sphere; AP will be its height, and PM the radius of its base. The convex surface of the cone, which we will call S, will be

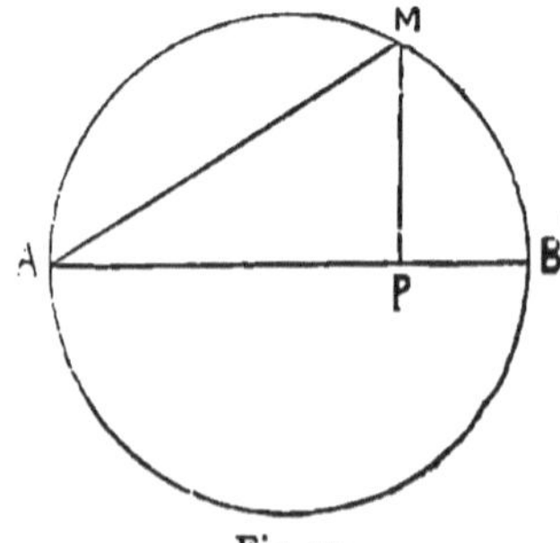

Fig. 10.

$$S = 2\pi PM \,.\, \tfrac{1}{2}AM = \pi PM \,.\, AM \qquad (1)$$

We have now to determine PM or AM, either of which will determine the other

Let

$$AB = 2a \text{ and } AP = x$$

then

$$\overline{PM}^2 = AP \,.\, PB = x(2a-x)$$

and

$$PM = \sqrt{x(2a-x)}$$

Again

$$AM = \sqrt{2ax}$$

Substituting these values in equation (1) we have

$$S = \pi PM \,.\, AM = \pi\sqrt{2ax - x^2} \,.\, \sqrt{2ax} = \pi\sqrt{4a^2x^2 - 2ax^3}$$

Differentiating this function we have

$$\frac{dS}{dx} = \frac{4a^2x - 3ax^2}{\sqrt{4a^2x^2 - 2ax^3}} = \frac{4a^2 - 3ax}{\sqrt{4a^2 - 2ax}} \qquad (2)$$

Placing this value of $\frac{dS}{dx}$ equal to zero we have

$$x = \frac{4a}{3}$$

In order to determine whether this value of x corresponds to a maximum or minimum of the function, it will be necessary to find the sign of the second differential coefficient.

Before doing this we will examine a method by which the operation may be somewhat abridged.

We have already seen that when the value of a function is reduced to zero by giving a particular value to the variable, it *does not follow* that its differential will also be reduced to zero by the same value of the variable (Art. 31), for the function in passing through zero may, and probably will, be passing from negative to positive, or *vice versa*, and, therefore, may have a differential or rate of change at that point, the same as at any other. Thus the latitude of a vessel on the equator is zero, but it may be changing as rapidly there as anywhere else.

Now if we wish to obtain the second differential coefficient for a particular value of the variable, we may take advantage of this circumstance. Suppose we find the first differential coefficient to be the product of two or more factors; either of these being reduced to zero will reduce the coefficient to zero. In this case we may obtain the value of the second differential coefficient for the corresponding value of the variable, without differentiating the entire coefficient. For suppose we have

$$\frac{du}{dx}=y\,y'\,y''$$

in which y . y' and y'' are functions of x; this product will be reduced to zero by any value of x that will reduce either factor to zero. Suppose that for $x=a$ we have $y=o$; if we differentiate the function we have

$$\frac{d\left(\frac{du}{dx}\right)}{dx}=\frac{d^2u}{dx^2}=\frac{d(y\,y'\,y'')}{dx}=\frac{y'\,y''\,dy}{dx}+\frac{y\,y''\,dy'}{dx}+\frac{y\,y'\,dy''}{dx}$$

and since $x=a$ reduces y to zero, the two last terms of this expression become equal to zero, and we have

$$\frac{d^2u}{dx^2}=\frac{y'\,y''\,dy}{dx}$$

hence to obtain the value of $\frac{d^2u}{dx^2}$ *for that value* of x that reduces one factor of the first differential coefficient to zero, we have only to multiply *its* differential coefficient by those factors which do *not* become zero, and then substitute the value of x. If for example we have

$$\frac{du}{dx}=x(x^2-a^2)=x(x+a)(x-a)$$

we may reduce it to zero by making

$$x=o,\ x=a \text{ or } x=-a$$

If we wish the value of $\frac{d^2u}{dx^2}$ for the first value of x, we have

$$\frac{d^2u}{dx^2}_{x=o}=x^2-a^2=-a^2$$

If for the second value we have

$$\frac{d^2u}{dx^2}_{x=a}=x(x+a)=2a^2$$

If for the third value we have

$$\frac{d^2u}{dx^2}_{x=-a}=x(x-a)=2a^2$$

Resuming now equation (2)

$$\frac{dS}{dx}=\frac{4a^2-3ax}{\sqrt{4a^2-2ax}}=\frac{a}{\sqrt{4a^2-2ax}}(4a-3x)$$

which becomes zero by making $3x=4a$ or $x=\frac{4a}{3}$, we shall have, for that value of x,

$$\frac{d^2S}{dx^2}=\frac{a}{\sqrt{4a^2-2ax}}\cdot\frac{d(4a-3x)}{dx}=\frac{-3a}{\sqrt{4a^2-2ax}}$$

which being negative shows that $x=\frac{4a}{3}$ corresponds to a maximum of the function.

Hence, *if a right cone be inscribed in a given sphere it will have the greatest possible convex surface when the axis of the cone is equal to two-thirds of the diameter of the sphere.*

Ex. 5. A point o (Fig. 11) being given within the right angle B A C, through which a line is to be drawn meeting the axes A B and A C, it is required to find the distance A n such that the length of the line between the points of its intersection with the axes shall be a minimum.

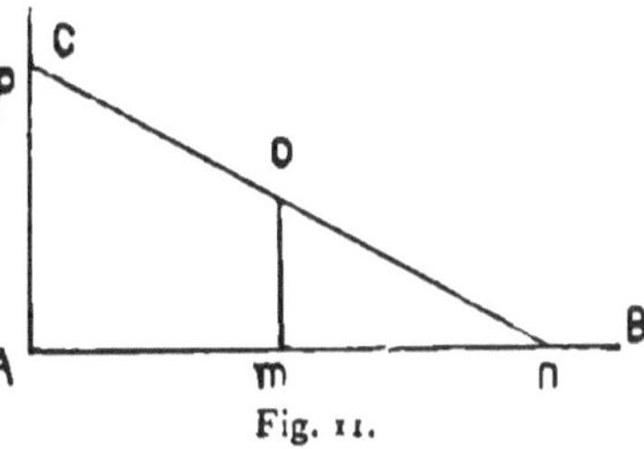

Fig. 11.

Let $Am=a$, $Om=b$ and $mn=x$. Then the right angled triangles Omn and pAn give the proportion $mn : Om :: An : Ap$ or

$$x : b :: a+x : Ap$$

whence

$$Ap=\frac{b(a+x)}{x} \text{ and } \overline{Ap^2}=\frac{b^2(a+x)^2}{x^2}$$

But

$$\overline{Ap}^2+\overline{An}^2=\overline{pn}^2$$

whence

$$\overline{pn}^2=\frac{b^2(a+x)^2}{x^2}+(a+x)^2$$

whence

$$pn=\frac{a+x}{x}\sqrt{b^2+x^2}$$

which is the function of which we are to find the minimum value.

Representing this function by u, and considering it as the product of two factors, we have

$$u=\frac{a+x}{x}\sqrt{b^2+x^2}$$

$$\frac{du}{dx}=\frac{a+x}{x}\cdot\frac{x}{\sqrt{b^2+x^2}}+\sqrt{b^2+x^2}\cdot\frac{-a}{x^2}$$

Reducing to a common denominator we have

$$\frac{du}{dx}=\frac{(a+x)x^2-a(b^2+x^2)}{x^2\sqrt{b^2+x^2}}=\frac{x^3-ab^2}{x^2\sqrt{b^2+x^2}}$$

Making this equal to zero we have

$$x=\sqrt[3]{ab^2}$$

To find if this is a minimum we differentiate again, but as the numerator of the differential coefficient is equal to zero for this value of x (Ex. 4), we multiply its differential by

$$\frac{1}{x^2\sqrt{b^2+x^2}}$$

which gives

$$\frac{d^2u}{dx^2}=\frac{3x^2}{x^2\sqrt{b^2+x^2}}=\frac{3}{\sqrt{b^2+x^2}}$$

a result that is essentially positive whatever may be the value of x. Hence $x=\sqrt[3]{ab^2}$ corresponds to a minimum length of the line pn. If a and b are equal, we have

$$x=b \text{ or } mn=Om$$

whence

$$Ap=An$$

Ex. 6. To find the maximum rectangle that can be inscribed in a given parabola.

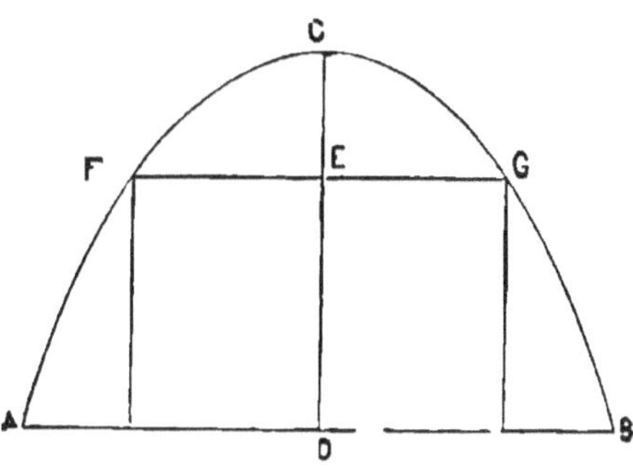

Fig. 12.

Let A C B (Fig. 12) be the parabola of which C D is the axis. Let D E be the height of the inscribed rectangle, and let $CD=a$ and $CE=x$, then $FE^2 = 2px$ and $FG=2\sqrt{2px}$

The area of the rectangle is

$$FG\times ED \text{ or } 2\sqrt{2px}(a-x)$$

which is to be a maximum.

Dropping the constant factor $2\sqrt{2p}$ (Ex. 2), representing the function by u and differentiating, we have

$$u=ax^{\frac{1}{2}}-x^{\frac{3}{2}} \text{ and } \frac{du}{dx}=\tfrac{1}{2}ax^{-\frac{1}{2}}-\tfrac{3}{2}x^{\frac{1}{2}}$$

which being made equal to zero gives

$$x=\frac{a}{3}$$

hence *the altitude of the rectangle is equal to two-thirds that of the parabola.*

To show that this is a maximum we differentiate again, and find

$$\frac{d^2u}{dx^2}=-\tfrac{1}{4}ax^{-\frac{3}{2}}-\tfrac{3}{4}x^{-\frac{1}{2}}$$

which is negative for every positive value of x, and therefore for $x=\frac{a}{3}$.

Ex. 7. What is the length of the axis of the maximum parabola that can be cut from a given right cone?

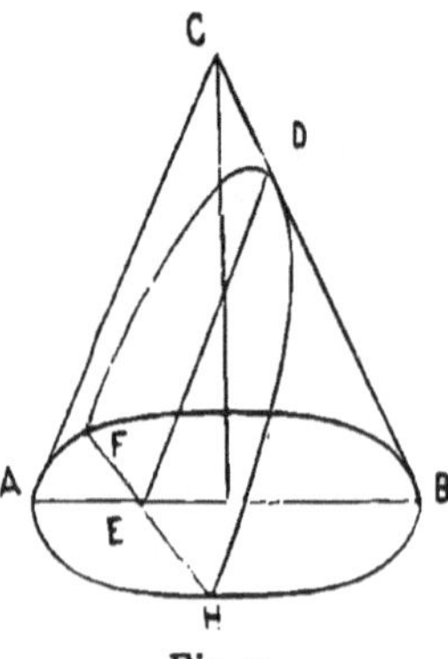

Fig. 13.

Let ABC (Fig. 13) be the given cone, and FDH the parabola cut from it. Let DE be the axis of the parabola, and AB the diameter of the base of the cone. Represent AB by a, AC by b, and BE by x. Then $AE=a-x$, $FE=\sqrt{ax-x^2}$. Also

$AB : AC :: EB : ED$ or $a : b :: x : ED$

hence

$$ED=\frac{bx}{a}$$

But the area of the parabola is equal to

$$\tfrac{2}{3}FH\times DE \text{ or } \tfrac{2}{3}\,\frac{xb}{a}\,.\,2\sqrt{ax-x^2}$$

which is, therefore, to be a maximum. Dropping the constant factor $\frac{4b}{3a}$ (Ex. 2), representing the function by u, and differentiating we have

$$u=\sqrt{ax^3-x^4} \tag{1}$$

$$\frac{du}{dx}=\frac{3ax^2-4x^3}{2\sqrt{ax^3-x^4}}=\frac{1}{2\sqrt{ax^3-x^4}}\,.\,(3ax^2-4x^3) \tag{2}$$

Placing the second member of equation (2) equal to zero we have $x=o$ and $x=\frac{3a}{4}$; and differentiating (2) for this last value of x we have (Ex. 4)

$$\frac{d^2u}{dx^2}=\frac{1}{2\sqrt{ax^3-x^4}}\,.\,(6ax-12x^2) \tag{3}$$

Substituting this value of x in the second member of this equation it is reduced to

$$\frac{d^2u}{dx^2}=-2\sqrt{3}$$

Hence the axis of the maximum parabola is three-fourths of slant height of the cone.

Ex. 8. It is required to determine the proportion of a cylinder, that shall have a given capacity, and whose entire surface shall be a minimum.

Let a^3 be the capacity of the cylinder, $x=\frac{1}{2}AB$ (Fig. 14) the radius of the base, and $y=AC$ be the height; then the two bases taken together will be equal to $2\pi x^2$, and the convex surface to $2\pi xy$, so that $2\pi(x^2+xy)$ is to be a minimum. Now $\pi x^2 y=a^3$, whence $y=\frac{a^3}{\pi x^2}$. Substituting this value of y, representing the function by u, and differentiating we have

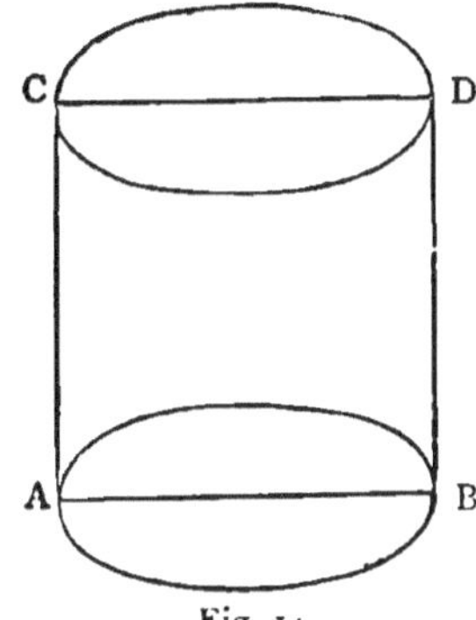

Fig. 14.

$$u=2\pi x^2+\frac{2a^3}{x} \qquad (1)$$

$$\frac{du}{dx}=4\pi x-\frac{2a^3}{x^2} \qquad (2)$$

$$\frac{d^2u}{dx^2}=4\pi+\frac{4a^3x}{x^4} \qquad (3)$$

Placing the value of $\frac{du}{dx}$ equal to zero we have

$$2\pi x^3=a^3=\pi x^2 y$$

whence

$$y=2x$$

or *the height of the cylinder is equal to the diameter of the base* and

$$x=\frac{a}{\sqrt[3]{2\pi}}$$

This value of x corresponds to a minimum value of the function, as is shown by substituting it in equation (3), which gives

$$\frac{d^2u}{dx^2}=12\pi$$

a positive quantity.

Ex. 9. To divide a right line into two parts such that one part multiplied by the cube of the other shall be a maximum.

Ans. The part cubed is three-fourths of the given line.

Ex. 10. To find the greatest right angled triangle that can be constructed on a given line as a hypothenuse.

Ans. The triangle must be isosceles.

Ex. 11. It is required to circumscribe about a given parabola, a minimum, isosceles triangle. What is the length of its axis?

Ans. Four-thirds the axis of the parabola.

Ex. 12. What is the altitude of the maximum cylinder that can be inscribed in a paraboloid?

Ans Half that of the paraboloid.

Ex. 13. The whole surface of a cylinder being given, how do the base and altitude compare with each other when the volume is a maximum? *Ans.*

Ex. 14. Required the minimum triangle formed by the axis, the produced ordinate of the extreme point, and the tangent to the curve of a parabola. *Ans.*

SECTION V.

APPLICATION OF THE DIFFERENTIAL CALCULUS TO THE THEORY OF CURVES.

SIGNIFICATION OF THE FIRST DIFFERENTIAL COEFFICIENT.

(34) In order to form such a conception of a line as will be adapted to the methods of the differential calculus, we must consider it to be *the path of a flowing point.*

The *law* which governs the movement of the point determines the nature of the line, as to form and position; and this law is expressed in the Cartesian system by the *equation* which shows the relation between the coördinates of the generating point in every position it may occupy throughout its movement.

The *direction* in which the point is moving is determined by the *relative rates* at which the coördinates are changing their values at the same moment. If the rate of change should be constantly the same in each of the coördinates, whether negative or positive, the generating point would move constantly in the same direction, describing a straight line; and this direction would be determined by the *ratio* of these rates, which in this case would be measured by the simultaneous increments or decrements of the coördinates themselves.

If, for instance, the coördinates AB and BC (Fig. 15) have each a constant rate of increase, the ratio of the increments CO and OC′ will be constant, and the generating point C will move in a straight line, whose direction will be determined by the *relative rates* with which those increments are produced; or since the rates, being uniform, may be represented by the simultaneous increments, the direction will be determined by the ratio $\frac{C'O}{CO}$.

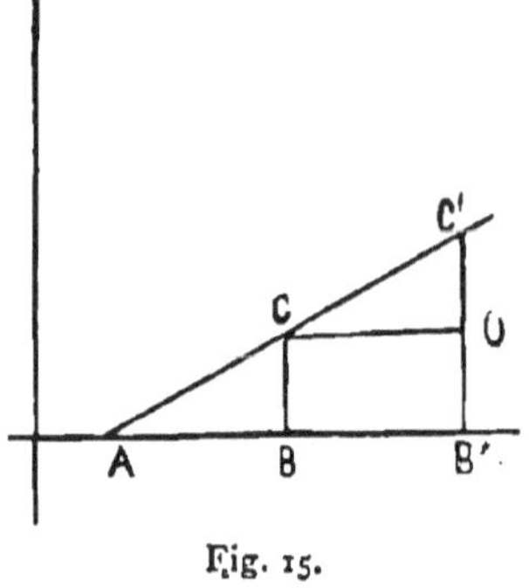

Fig. 15.

But if, while the rate of change in one of the coördinates is constant, that of the other should constantly vary, the ratio of their simultaneous increments would be constantly changing and the point would describe a curve, whose character would be determined by the *law* which should govern these varying rates of change; and *this law would be expressed in the equation of the curve.* But if the varying rate of change in the coördinate should, at any point in the curve, cease to vary, and should continue afterwards constantly the same as at that point, the generating point would cease to describe a curve, and would move in a straight line in the direction to which it was *then tending;* and this direction would be determined by the ratio between the rates of change in the coördinates as they existed *at the instant they both became uniform.*

Now the tangent to any curve is the line which would be described by the generating point if it were to move in the direction to which it is tending on its arrival at the point of tangency; just as a stone, when it leaves a sling, describes a line tangent to the curve in which it was moving at the instant. Hence the ratio of the rates of change in the coördinates of any point of a curve determines the direction of the line tangent to the curve at that point.

Suppose, for example, that AB (Fig. 16) has a uniform rate of increase, while BC has a rate of increase that is constantly diminishing; the point C would describe a curve.

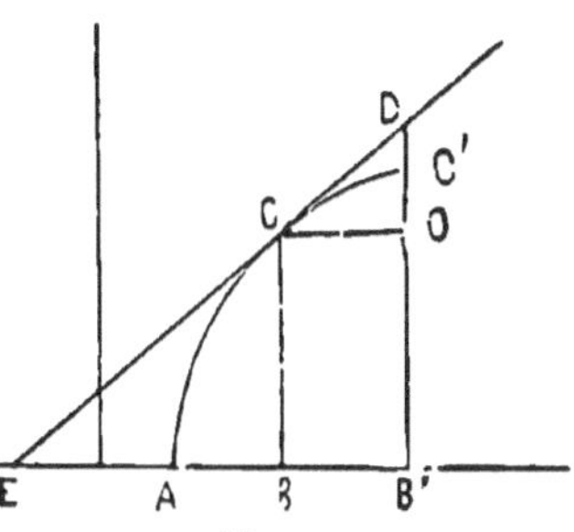

Fig. 16.

Now let us suppose that the generating point on arriving at the point C of the curve should continue to move in the direction toward which it was then tending, and should with a uniform motion, at the same rate as it had at C, describe the right line CD, then this line would be tangent to the curve at the point C. From D draw a line DB′ parallel to the ordinate, and meeting the axis of abscissas at B′; and from C draw a line parallel to the axis of abscissas meeting DB′ in O. The triangle CDO will have important properties which will require careful investigation.

From whatever point in the line CD the line DB′ is drawn, the ratio between the lines CD, CO and DO will be the same, and hence as CD is described at a uniform rate, equal to that with which the generating point is moving at the point C, the lines CO and DO will also be described at a uniform rate equal to that with which AB and BC were increasing at the same instant. Hence the three lines CD, CO and OD are the increments that *would take place* in the *arc*, the *abscissa* and the *ordinate* in the same unit of time, at their several rates of change existing when the generating point of the curve is at C; and, therefore, CO and OD may be taken as *symbols* representing the *rate* of increase of the abscissa and ordinate of the curve, while CD will represent the rate of increase of the curve itself at the point C; and is at the same time tangent to it at that point.

So that if we designate the length of the curve by s, and consider CD as representing ds, we shall have CO$=dx$ and OD$=dy$ for the point C of the curve.

The tangent of the angle which the tangent line CD makes with the axis of abscissas is equal to $\frac{DO}{CO}$, and hence calling this angle v.

$$\frac{dy}{dx}=\text{tang. } v \tag{1}$$

and since $\overline{CD}^2=\overline{CO}^2+\overline{OD}^2$ we have

$$ds^2=dx^2+dy^2 \tag{2}$$

We shall have frequent occasion to use these two equations in investigating the properties of curves.

(35) The usual method of obtaining equation (1) is, to suppose an *actual* increment BB′ (Fig. 16) given to the abscissa, and to find the corresponding increment C′O of the ordinate from the equation of the curve. The ratio of these two increments will not give the tangent of the angle v, which is equal to $\frac{DO}{CO}$, but will approach it as the increments decrease, and when they become infinitesimal or vanish, there is no difference between $\frac{DO}{CO}$ and $\frac{C'O}{CO}$.

This manner of reasoning we have already discussed in the introduction to this work, and its defects have been shown. We commit an error in giving an actual increment to the abscissa and ordinate, for their rates of increase are not obtained from any actual increase in value (except where the rate is uniform), but from the *law* of change derived from the equation of the curve; and the suppositive increments which we give are not real, but symbolical, representing what they *would be*, by the operation of the *law* controlling them at that instant, and are, therefore, a symbolical expression of that law.

The truth is that CO and OD, so far from being infinitely small may have *any value whatever* assigned to them; so that we may consider them as *variables* whose simultaneous values always correspond to some point in the tangent line. In fact, if we differentiate the equation of a curve, and consider x and y as constants for the point of tangency, dx and dy may be considered as the variable coördinates of the tangent line, with the origin at the point of tangency, and the axes parallel to the primitive axes. Under these conditions the differential equation of the curve becomes the equation of its tangent line. This can easily be shown by a few examples.

Ex. 1. The equation of the circle with the origin at A (Fig. 17) is

$$y^2=2Rx-x^2$$

which being differentiated becomes

$$ydy=(R-x)dx$$

If now we consider x and y as constants for the point P, and $dx(=PE)$ and $dy(=ET)$ as variables, we will replace the first by m and n, and the latter by x and y, and we have

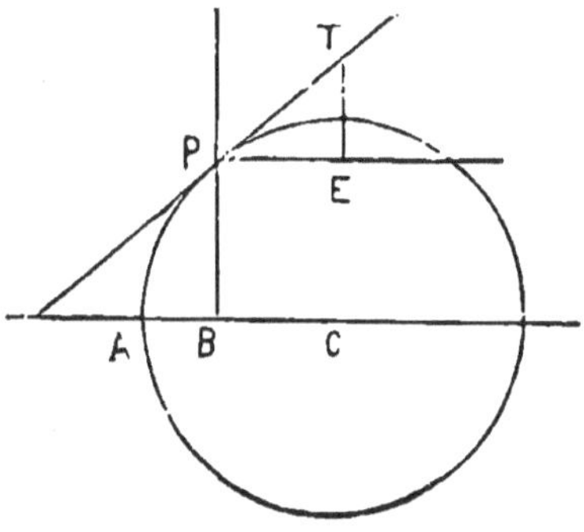

Fig. 17.

$$my=(R-n)x \qquad (1)$$

which is the equation of the tangent line PT with the origin at P, while PE and TE are the abscissa and ordinate of the line, and m and n are coördinates of the new origin referred to the primitive one, or AB and BP.

Suppose now we transfer the origin to C, the center of the circle. The formulas for transferring to a new origin in a system of parallel axes are

$$y=b+y' \text{ and } x=a+x'$$

where a and b are the coördinates of the new origin. In this case a is equal to $BC=AC-AB=R-n$, and b is equal to $PB=-m$, and hence by substitution,

$$m(-m+y')=(R-n)(R-n+x') \qquad (2)$$

Calling x'' and y'' the coördinates of the point of tangency for the new origin, we have

$$x''=-R+n \text{ or } R-n=-x''$$

also

$$y''=BP=m$$

Substituting these values in equation (2) we have

$$y''(-y''+y')=-x''(-x''+x')$$

whence

$$y'\,y''+x'\,x''=x''^2+y''^2=R^2$$

or dropping the accents

$$yy''+xx''=R^2$$

which is the equation of the tangent to the circle, the origin being at the center and x'' and y'' the coördinates of the point of tangency.

Ex. 2. If we differentiate the equation of the ellipse referred to its center and axes, we have

$$A^2ydy+B^2xdx=0$$

Making

$y(=BP)$ and $x(=OB)$

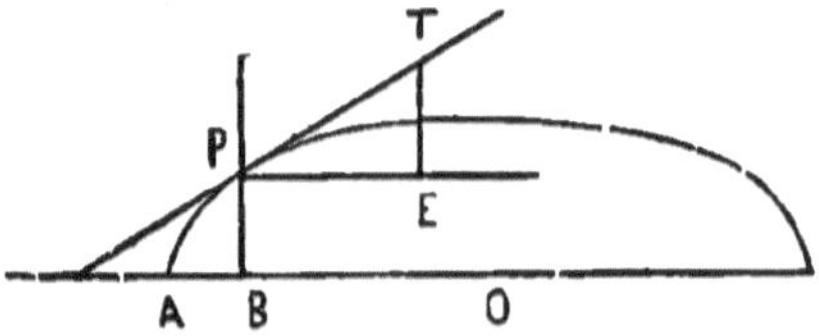

Fig. 18.

(Fig. 18) constant and $dy(=TE)$ and $dx(=PE)$ variables, and replacing y by y'' and x by x'', dx by x and dy by y, we have

$$A^2y''y+B^2x''x=0$$

which is the equation of the tangent line to the ellipse, with the origin at P, the point of tangency, and the axes parallel to the primitive ones; x'' and y'' representing the coördinates of the new origin referred to the primitive one, and x and y the variable coördinates of the tangent line referred to the new origin.

If we transfer the origin back to the primitive one we shall have

$$x=a+x' \text{ and } y=b+y'$$

where a and b are the coördinates of the new origin — that is, the center of the ellipse. But $a=-x''$ and $b=-y''$ (for a is essentially positive while x'' is essentially negative, and b is essentially negative while y'' is essentially positive), and substituting these values for x and y we have

$$A^2y''y'+B^2x''x'=A^2y''^2+B^2x''^2=A^2B^2$$

or dropping the accents

$$A^2y''y+B^2x''x=A^2B^2$$

Ex. 3. Differentiating the equation of the parabola we have

$$ydy=pdx$$

Representing x, y, dx and dy by x'', y'', x and y respectively, we have by substitution

$$y''y=px$$

for the equation of the tangent line to the parabola, with the origin at the point of tangency. If we transfer it to the vertex by making $y=b+y'$ and $x=a+x'$, in which $b=-y''$ and $a=-x''$, we have

$$-y''^2+y'y''=px'-px''$$

in which x'' and y'' are the coördinates of the point of tangency for the origin at A. Hence $y''^2=2px''$, and substituting we have

$$-2px''+y'y''=px'-px''$$

or, dropping the accents,

$$yy''=px+px''=p(x+x'')$$

which is the equation of the tangent to the parabola at the point whose coördinates are x'' and y'', the origin being at the vertex.

Ex. 4. Lastly we will take the equation of the Hyperbola referred to its center and asymptotes

$$xy=\frac{A^2+B^2}{4}$$

from which

$$xdy+ydx=0$$

Replacing y=PB (Fig. 19) by y'', x=AB by x'', dy=ET by y and dx=EP by x, we have

$$y''x+x''y=0$$

in which x'' and y'' are the coördinates of the new origin referred to the primitive one, and x and y are the variable coördinates of the tangent line TP; the origin being at P and the coördinate axes, PM and PN, parallel to the asymptotes.

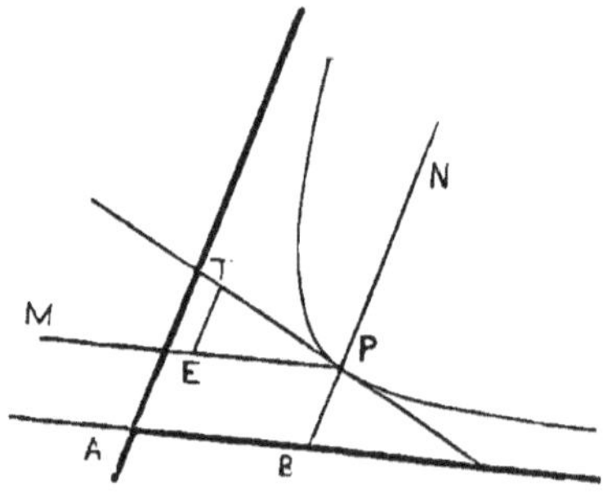

Fig. 19.

To transfer the origin to A the center of the hyperbola make $y=b+y'$, and $x=a+x'$; b being equal to $-y''$ and $a=-x''$; and substitute these values for x and y. This gives

$$y''(x'-x'')+x''(y'-y'')=0$$

or, dropping the accents

$$y-y''=-\frac{y''}{x''}(x-x'') \text{ or } yx''+xy''=\frac{A^2+B^2}{2}$$

which is the equation of the tangent line to the hyperbola referred to its center and asymptotes.

Thus it clearly appears that differentials are not infinitely small quantities, but are *symbols* to express the *rates* or *laws of variation*, which are, in fact, VARIABLE FUNCTIONS OF THE GIVEN VARIABLES.

SIGN OF THE FIRST DIFFERENTIAL COEFFICIENT

(36) If x and y represent the coördinates of any curve, and while x increases uniformly y should have a positive value and, also, *increase*, its differential will be positive and the curve will tend to *leave* the axis of abscissas in a positive direction; but if y should be *decreasing* while its value is positive, its differential will be negative and the curve will *approach* the axis of abscissas on the positive side.

Again if y has a negative value and *increasing*, its differential will be *negative*, and the curve will be *receding* from the axis of abscissas on the negative side; while if it is decreasing (being still negative) its differential will be *positive*, and the curve will be *approaching* the axis of abscissas on the negative side (see definition of a differential, Art. 3).

Hence the following rule:

When the ordinate and its first differential have the SAME *sign the curve is receding from the axis of abscissas, and when they have* DIFFERENT SIGNS *the curve is approaching that axis.*

NOTE.—The differential of the independent variable is supposed to be constant and *positive*, and hence the sign of the *differential coefficient* is the same as that of the differential itself of the function.

This rule may be illustrated by means of the circle (Fig. 20) whose equation (the origin being at A) is

$$y^2=2Rx-x^2$$

from which we obtain

$$dy=\frac{R-x}{y}dx$$

We see here that from A to C, y and its differential have the same sign, and the curve recedes from the axis of abscissas. The same is true of the curve from A to D. But from C to B, and from D to B, where x is greater than R, the sign of y will be contrary to that of dy, and the curve approaches the axis of abscissas on both sides.

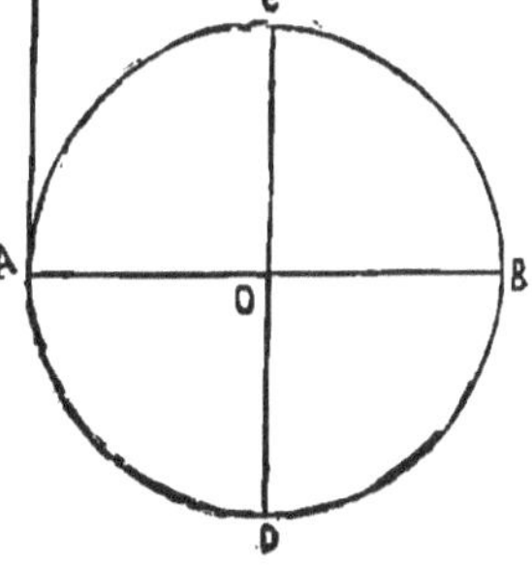

Fig. 20.

We arrive at the same result if we consider $\frac{dy}{dx}$ as representing the tangent of the angle which the tangent line makes with the axis of abscissas. From A to C and from A to D, where the curve leaves the axis of abscissas, the sign of the tangent and of y are alike; while from C to B and from D to B their signs are contrary.

At the point A where $y=o$, the value of $\frac{dy}{dx}=\frac{R-x}{y}$ becomes infinite, and the curve departs at right angles from the axis of abscissas. While at the points C and D the value $\frac{dy}{dx}$ becomes zero, and the curve neither approaches nor recedes from the axis of abscissas. And this corresponds with the value of the tangent of the angle made by the tangent line with the axis of abscissas; at A and B, $\frac{dy}{dx}=\infty$ and the angle is a right angle; at C and D, $\frac{dy}{dx}=0$ and the angle is zero; the tangent line being parallel to the axis of abscissas.

SECTION VI.

DIFFERENTIALS OF TRANSCENDENTAL FUNCTIONS.

PROPOSITION I.

(37) To find the differential of a constant quantity raised to a power having a variable exponent.

Let the constant quantity be represented by a and the variable exponent by v; then the function will be

$$a^v$$

If we add an increment to v which we will call m we shall have

$$a^{v+m}=a^v a^m \qquad (1)$$

Differentiating this equation we have

$$da^{v+m}=a^m da^v \qquad (2)$$

and dividing equation (2) by equation (1) we have

$$\frac{da^{v+m}}{a^{v+m}}=\frac{da^v}{a^v} \qquad (3)$$

This equation being true, irrespective of any particular value of m, it will be true for any value we may assign to it; hence the differential of a constant quantity raised to a power denoted by a variable exponent, divided by the power itself is a constant quantity, or

$$\frac{da^v}{a^v}=C$$

But we have seen (Art. 6) that the differential of the variable is always a factor in the differential of the function. Hence dv will be a factor of C. Calling the other factor k we have

$$C=kdv$$

whence

$$\frac{da^v}{a^v}=kdv$$

or

$$da^v=a^v kdv$$

The problem now is to find the value of k. For this purpose we expand a^v by Maclaurin's theorem and have

$$a^v=A+Bv+Cv^2+Dv^3+Ev^4+\text{etc} \qquad (1)$$

in which

$$A=a^v,\ B=\frac{da^v}{dv},\ C=\frac{d^2a^v}{2dv^2},\ D=\frac{d^3a^v}{2\cdot 3\cdot dv^3},\ \text{etc.}$$

when v is made equal to zero.

But we have found

$$\frac{da^v}{dv}=a^v k$$

hence

$$d\left(\frac{da^v}{dv}\right)=da^v k=a^v k^2 dv$$

or

$$\frac{d^2a^v}{dv^2}=a^v k^2$$

and similarly

$$\frac{d^3a^v}{dv^3}=a^v k^3,\ \frac{d^4a^v}{dv^4}=a^v k^4$$

from which making $v=o$ we have

$$A=1,\ B=k,\ C=\frac{k^2}{2},\ D=\frac{k^3}{2\cdot 3},\ E=\frac{k^4}{2\cdot 3\cdot 4},\ \text{etc.}$$

Substituting these values in equation (1) we have

$$a^v=1+kv+\frac{k^2v^2}{2}+\frac{k^3v^3}{2\cdot 3}+\frac{k^4v^4}{2\cdot 3\cdot 4}+\text{etc.}$$

and making $v=\frac{1}{k}$ this becomes

$$a^{\frac{1}{k}}=1+1+\tfrac{1}{2}+\tfrac{1}{2.3}+\tfrac{1}{2.3.4}+\text{etc.}=2.718282+$$

If we represent this number by e we have

$$a^{\frac{1}{k}}=e \text{ or } a=e^k$$

If e is made the base of a system of logarithms k would be the logarithm of a to that base.

This was done by Napier, the inventor of logarithms, and the system having that base is called the Naperian system. We shall indicate the logarithms of that system by the notation *log.*, while the logarithms of other systems will be noted by *Log.* We have, therefore,

$$da^v=a^v \log. adv.$$

that is,

The differential of a quantity raised to a power denoted by a variable exponent, is equal to the power multiplied by the Naperian logarithm of the constant quantity into the differential of the exponent.

Proposition II.

(38) To find the differential of the logarithm of a variable quantity.

Let the quantity be represented by r and its logarithm by v, the base of the system being represented by a. Then we have

$$r=a^v \text{ and } dr=a^v kdv$$

whence

$$dv=\frac{dr}{a^v k} \text{ or } d\,Log.\ r=\frac{1}{k}\cdot\frac{dr}{r}$$

Representing $\frac{1}{k}$ by M we have

$$d\,Log.\ r=M\frac{dr}{r}$$

in which M is the reciprocal of the Naperian logarithm of the base a, and is called the *modulus* of the system of logarithms of which a is the base. Hence

The differential of the logarithm of a variable quantity is equal to the modulus of the system to which the logarithm belongs, into the differential of the quantity divided by the quantity.

In the Naperian system the modulus is, of course, *one.* Hence in that system

$$d\ log.\ r = \frac{dr}{r}$$

from which we learn that in the Naperian system the rate of increase of the natural number, whatever may be its value, divided by the rate of increase of its logarithm, is always equal to the number itself.

Note.—This principle was used by Napier himself in constructing his table of logarithms, and explains his selection of his peculiar base. Hence he is one of the first discoverers of the *principle* of the differential calculus, although he never applied it otherwise than to logarithms.

(39) If we call e the base of the Naperian system of logarithms, a the base of any other system, m the logarithm of p to that base, n the Naperian logarithm of p, and s the Naperian logarithm of a we shall have

$$p = a^m \quad p = e^n \quad a = e^s$$

hence

$$p = e^{sm} = e^n$$

wherefore

$$n = sm \text{ or } m = \frac{n}{s} = \frac{1}{s} n$$

but $\frac{1}{s}$ is the modulus of the system, and hence

The logarithm of a number in any system is equal to the Naperian logarithm of that number multiplied by the modulus of the system.

This property is not peculiar to the Naperian system. The logarithm of a number in *any* system is equal to the logarithm of the same number in the common system multiplied by the reciprocal of the common logarithm of the base of the new system.

In fact, in any two different systems, the ratio between the logarithms of the same number is constant. Thus let a and b be the bases of two systems, m and n two numbers, and x and y their logarithms in the first, and u and v their logarithms in the second system; then

$$m=a^x \quad n=a^y \quad m=b^u \quad n=b^v$$

whence

$$a^x=b^u \text{ and } a^y=b^v$$

whence

$$a=b^{\frac{u}{x}} \text{ and } a=b^{\frac{v}{y}}$$

whence

$$\frac{u}{x}=\frac{v}{y}$$

or the ratio between the logarithms of the same number in two different systems is constant and equal to the ratio between the logarithms of any other number taken in the same systems. Hence

$$log.\ a : com.\ Log.\ a :: log.\ 10 : com. Log.\ 10=1$$

whence

$$com.\ Log.\ a=\frac{log.\ a}{log.\ 10}=M\ log.\ a$$

as we have seen.

Proposition III.

(40) To find the differential of the sine of an arc.

Let APD (Fig. 21) be a circle whose center is at O. Let POA be the given angle, then PB will be the sine of the arc AP, and also an ordinate of the circle to the axes OX and OY; while OB will be the cosine of the same angle, and also the abscissa of the point P of the curve, and AB the versed sine of the angle.

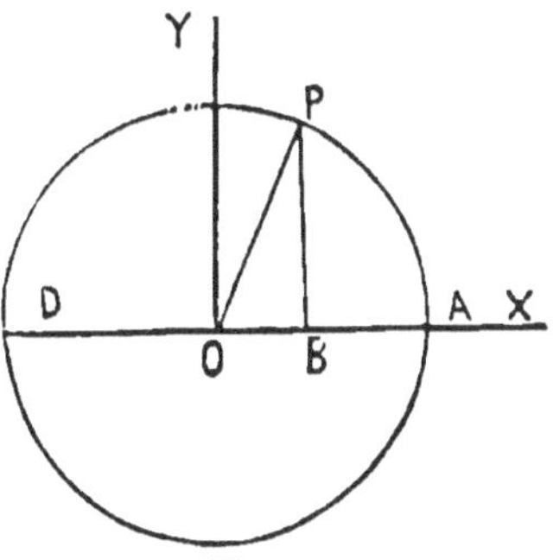

Fig. 21.

From the equation of the circle with the origin at the center we obtain

$$xdx = -ydy$$

and

$$dx^2 = \frac{y^2 dy^2}{x^2}$$

If we represent the arc AP by s we have (Art. 34)

$$ds^2 = dx^2 + dy^2$$

whence

$$ds^2 = \frac{y^2 dy^2}{x^2} + dy^2 = \frac{(x^2 + y^2)}{x^2} dy^2$$

But

$$x^2 + y^2 = R^2$$

whence

$$ds^2 = \frac{R^2 dy^2}{x^2}$$

from which we obtain

$$dy = \frac{x}{R} ds$$

or

$$d \sin. s = \frac{\cos. s}{R} ds \qquad (1)$$

that' is,

The differential of the sine of an arc is equal to the cosine of the arc into the differential of the arc divided by radius.

Proposition IV.

(41) To find the differential of the cosine of an arc.

From the equation

$$\cos. s = \sqrt{R^2 - \sin.^2 s}$$

we have

$$d \cos. s = \frac{-\sin. s . d \sin. s}{\sqrt{R^2 - \sin.^2 s}}$$

Substituting the value of $d \sin. s$ (Art. 40) and replacing the denominator by $\cos. s$, we have

$$d\cos. s = \frac{-\sin. s \cos. s \,.\, ds}{\text{R} \cos. s} = \frac{-\sin. s}{\text{R}} ds \qquad (2)$$

that is,

The differential of the cosine of an arc is equal to minus the sine of the arc into the differential of the arc, divided by radius.

PROPOSITION V.

(42) To find the differential of the tangent of an arc.

From the equation

$$\text{tang. } s = \frac{\text{R} \sin. s}{\cos. s}$$

we obtain (Art. 14)

$$d\,\text{tang. } s = \frac{\text{R} \cos. s \,.\, d\sin. s - \text{R} \sin. s \,.\, d\cos. s}{\cos.^2 s}$$

Substituting for $d\sin. s$ and $d\cos. s$ their values (Art. 40 and 41) we have

$$d\,\text{tang. } s = \frac{\text{R} \cos.^2 s . ds + \text{R} \sin.^2 s . ds}{\text{R} \cos.^2 s}$$

or

$$d\,\text{tang. } s = \frac{(\cos.^2 s + \sin.^2 s) ds}{\cos.^2 s} = \frac{\text{R}^2}{\cos.^2 s} ds \qquad (3)$$

that is

The differential of the tangent of an arc is equal to the square of the radius into the differential of the arc divided by the square of the cosine.

PROPOSITION VI.

(43) To find the differential of the cotangent of an arc.

From the equation

$$\cot. s = \frac{\text{R}^2}{\text{tang. } s}$$

we obtain

$$d\cot. s = \frac{-\text{R}^2 d\,\text{tang. } s}{\text{tang.}^2 s}$$

Substituting for $d\,\text{tang.}\,s$ its value from equation (3) we have

$$d\,\text{cot.}\,s=\frac{-R^4 ds}{\text{cos.}^2 s\ \text{tang.}^2 s}=\frac{-R^2}{\text{sin.}^2 s}ds \tag{4}$$

that is

The differential of the cotangent of an arc is negative, and equal to the differential of the arc multiplied by the square of the radius, and divided by the square of the sine.

Proposition VII.

(44) To find the differential of the secant of an arc.

From the equation

$$\text{sec.}\,s=\frac{R^2}{\text{cos.}\,s}$$

we have

$$d\,\text{sec.}\,s=\frac{-R^2\,d\,\text{cos.}\,s}{\text{cos.}^2 s}$$

Substituting the value of $d\,\text{cos.}\,s$ from equation (2) we have

$$d\,\text{sec.}\,s=\frac{R\ \text{sin.}\,s}{\text{cos.}^2 s}ds \tag{5}$$

that is

The differential of the secant of an arc is equal to the differential of the arc multiplied by the radius into the sine, divided by the square of the cosine.

Proposition VIII.

(45) To find the differential of the cosecant of an arc.

From the equation

$$\text{cosec.}\,s=\frac{R^2}{\text{sin.}\,s}$$

we obtain

$$d\,\text{cosec.}\,s=\frac{-R^2\,d\,\text{sin.}\,s}{\text{sin.}^2 s}$$

Substituting the value of d sin. s from equation (1) we have

$$d \text{ cosec. } s = \frac{-\text{R cos. } s}{\text{sin.}^2 s} ds \qquad (6)$$

that is

The differential of the cosecant of an arc is equal to minus the differential of the arc multiplied by radius into the cosine, divided by the square of the sine.

Proposition IX.

(46) To find the differential of the versed sine of an arc.

From the equation

$$\text{ver. sin. } s = \text{R} - \text{cos. } s$$

we have

$$d \text{ ver. sin. } s = -d \text{ cos. } s$$

Substituting for d cos. s its value from equation (2) we have

$$d \text{ ver. sin. } s = \frac{\text{sin. } s}{\text{R}} ds \qquad (7)$$

that is,

The differential of the versed sine of an arc is equal to the differential of the arc multiplied by the sine and divided by radius.

(47) In these equations the arc is supposed to be the independent variable; and the generating point is supposed to flow around the circumference at a uniform rate.

The differential of the arc may easily be found, considering it as a dependent variable and either the sine, cosine, tangent, etc., as the independent one varying uniformly.

If we take the sine as the independent variable we have from equation (1)

$$ds = \frac{\text{R}}{\text{cos. } s} d \text{ sin. } s = \frac{\text{R}}{\sqrt{\text{R}^2 - \text{sin.}^2 s}} d \text{ sin. s} \qquad (8)$$

If we take the cosine we have from (2)

$$ds = -\frac{\text{R}}{\text{sin. } s} d \text{ cos. } s = -\frac{\text{R}}{\sqrt{\text{R}^2 - \text{cos.}^2 s}} d \text{ cos. } s \qquad (9)$$

If we take the tangent we have from (3)

$$ds=\frac{\cos.^2 s}{R^2}d\tan g.\, s=\frac{R^2}{\sec.^2 s}d\tan g.\, s$$

but

$$\sec.^2 s=R^2+\tan g.^2 s$$

hence

$$ds=\frac{R^2\, d\tan g. s}{R^2+\tan g.^2 s} \qquad (10)$$

Lastly, if we take the versed sine we have from equation (7)

$$ds=\frac{R}{\sin.\ s}d\text{ ver. sin. } s$$

but since

$$\sin.\ s=\sqrt{(2R-\text{ver. sin. } s)\text{ ver. sin. } s}$$

we have

$$ds=\frac{R\, d\text{ ver. sin. } s}{\sqrt{(2R-\text{ver. sin. } s)\text{ ver. sin. } s}} \qquad (11)$$

If, in equations (8), (9), (10) and (11), we represent sin. s by u, cos. s by x, tang. s by y, and ver. sin. s by z, and consider $R=1$, we shall have

$$ds=\frac{du}{\sqrt{1-u^2}} \qquad (12)$$

$$ds=-\frac{dx}{\sqrt{1-x^2}} \qquad (13)$$

$$ds=\frac{dy}{1+y^2} \qquad (14)$$

$$ds=\frac{dz}{\sqrt{2z-z^2}} \qquad (15)$$

From these equations we can find the rate of change in the arc when we know that of either of the four trigonometrical lines.

SIGNIFICATION OF THE DIFFERENTIAL EQUATIONS OF THE TRIGONOMETRICAL LINES.

(48) Describe the circle ADBE (Fig. 22) from the center O. Draw the diameter AB and the tangent TT′ at its extremity. Let sc, $s'c'$, $s''c''$, and $s'''c'''$ be the sines of the arcs As, ADs', ADBs'' and ADBEs''', and OT and OT' (negative for ADs' and ADBs'') be the secants of the same arcs; then AT will be the tangent of the arcs As and ADBs'', and AT' will be the tangent of the arcs ADs' and ADBEs'''; oc, oc', oc'' and oc''' will be the cosines of the same arcs, and Ac, Ac', Ac'' and Ac''' will be their versed sines.

Suppose now the generating point of the circle to move from A around through D, B and E back to A with a uniform rate of motion, then

From equation (1) we find that at the beginning where cos. = R the rate of increase of the sine is the same as that of the arc and is positive. As the sine increases the rate immediately begins to decrease, being always in proportion to the cosine, until the generating point arrives at D and the cosine becomes zero. The sine then ceases to increase, being at a maximum, and its rate of increase is zero.

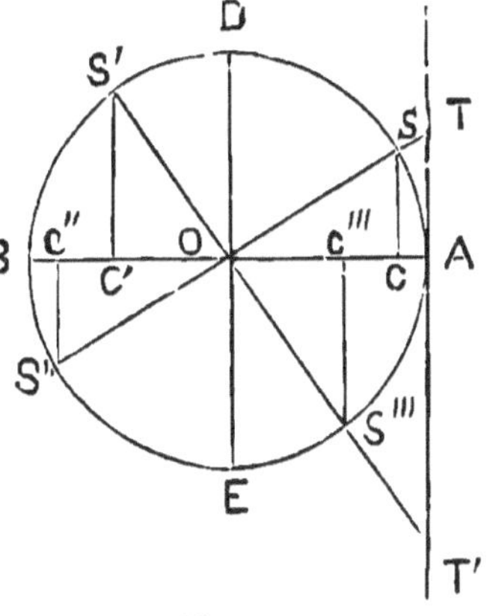

Fig. 22.

In the second quadrant, the sine although still positive, decreases, and hence its rate of change is negative as shown by the cos. s, which is negative in that quadrant, at the point B the rate of decrease has become equal to the rate of increase of the arc, although the value of the sine itself has become zero, and hence at that point d sin. $s = -ds$. In the third quadrant the sine is negative and increasing, hence its rate is also negative, as is shown by the cosine which is negative. At the point E the negative increase ceases and the rate becomes zero as does the value of the cosine. In the fourth quadrant, while the sine is negative, it is diminishing, and hence its rate of change is positive which is also indicated by the cosine which is positive for that quadrant.

From equation (2) we learn that in the first and second quadrants where the sine is positive, the rate of change in the cosine is negative, and in the third and fourth quadrants where the sine is negative the rate of change in the cosine is posi ive. We learn the same thing from the figure, for from A to B the cosine decreases, being positive, or increases, being negative; while from B to A, for the third and fourth quadrants, it decreases, being negative, or increases, being positive; the rate of change being at all times in direct proportion to the value of the sine.

From equation (3) we learn that the rate of change in the tangent is at all times positive, and we learn the same thing from the figure, for as the arc increases from any point whatever, the extremity of the secant which limits the tangent will move *upward* in a positive direction, and the tangent will increase in the first and third angles from A to positive infinity, and decrease in the second and fourth from negative infinity to A; so that it will always increase positively or decrease negatively, and hence its rate of change is always positive (Art. 3). At A and B the rate is the same as that of the arc, and of the sine, being equal to ds, while at D and E it is infinite.

Similarly we learn from equation (4) that the rate of change in the cotangent is always negative, as it either decreases being positive, or increases being negative, for all points of the circle.

From equation (5) we learn that the rate of change in the secant has the same sign as the sine, and hence is positive in the first and second quadrants, and negative in the third and fourth. By inspecting the figure we see that in the first quadrant the positive secant OT increases; in the second, the negative secant OT′ decreases; in the third, the negative secant OT increases; and in the fourth, the positive secant OT′ decreases. At the point A the rate of increase of the secant is zero for sin. $s=o$, while the secant itself equals R (a minimum). We see also that the rate of the secant is equal to that of the tangent multiplied by the sine; and since the sine is (except at two points) always less than 1, the tangent increases faster than the secant until the arc equals 90°, when the sine is 1 and the rates become equal being infinite.

Equation (6) will give a similar result for the cosecant in connection with the cosine and cotangent.

We learn from equation (7) that the rate of increase of the versed sine corresponds at all times to the value of the sine, and is therefore positive in the first and second quadrants, and negative in the third and fourth. The figure shows that the versed sine increases positively from A to B, the rate of increase being an increasing one (corresponding to the sine) in the first quadrant, and then a decreasing one in the second, but still positive. At B it ceases to increase and begins to decrease, first at an increasing rate in the third quadrant, and then at a decreasing rate in the fourth, until at A the versed sine and the rate both become zero. Hence in the two last quadrants the rate is negative corresponding to the sine, while the versed sine is *always positive*. At A and B the rate of change is nothing, as it should be, since at those points the generating point of the circle tends to move in a direction perpendicular to the line on which the versed sine is laid off, and, therefore, does not tend to alter its value; while at D and E, the generating point moves parallel to the line of the versed sine, and, therefore, at those points they should have the same rate, and this we find to be the case, for at D

$$d \text{ ver. sin. } s = \frac{\text{sin. } s}{\text{R}} ds = ds$$

and at E

$$d \text{ ver. sin. } s = \frac{\text{sin. } s}{R} ds = -ds$$

because sin. s at that point is negative.

VALUES OF TRIGONOMETRICAL LINES.

(49) We are enabled by Maclaurin's theorem to develop the sine and cosine of an arc in terms of the arc itself; for let s be the arc and u its sine, we shall have (Art. 24)

$$u = A + Bs + Cs^2 + Ds^3 + \text{ etc.}$$

and (Art. 40, 41) making $R=1$ we have

$$\frac{du}{ds} = \text{cos. s}, \quad \frac{d^2 u}{ds^2} = -\text{sin. } s, \quad \frac{d^3 u}{ds^3} = -\text{cos. } s, \text{ etc.}$$

making $s=0$ we have

$$(u)=0, \quad \left(\frac{du}{ds}\right)=1, \quad \left(\frac{d^2u}{ds^2}\right)=0, \quad \frac{d^3u}{ds^3}=-1, \text{ etc.}$$

whence

$$u = \text{sin. } s = s - \frac{s^3}{2 \cdot 3} + \frac{s^5}{2 \cdot 3 \cdot 4 \cdot 5} -, \text{ etc.}$$

If we represent by u the cosine s then

$$\frac{du}{ds} = -\text{sin. } s, \quad \frac{d^2 u}{ds^2} = -\text{cos. } s, \quad \frac{d^3 u}{ds^3} = \text{sin. } s, \text{ etc.}$$

Making $s=0$ we have

$$(u)=1, \quad \left(\frac{du}{ds}\right)=0, \quad \frac{d^2 u}{ds^2} = -1, \quad \frac{d^3 u}{ds^3} = 0, \text{ etc.}$$

whence

$$u = \text{cos. } s = 1 - \frac{s^2}{2} + \frac{s^4}{2 \cdot 3 \cdot 4} - \frac{s^6}{2 \cdot 3 \cdot 4 \cdot 5 \cdot 6} +, \text{etc.}$$

These series are very converging, and for small arcs will give the length of the sine and cosine quite accurately.

In order to apply these formulas, we take the length of a quadrant, which is $\frac{\pi}{2}$, the radius being 1; and this divided by 90 and then by 60, will give the length of one minute of arc, from which we can obtain the length of any number of minutes or degrees. Substituting the value of the arc thus found in the formulas, we obtain the length of the natural sine or cosine.

If we wish these values for any other radius we shall have for sin. $s=u$

$$\frac{du}{ds}=\frac{\cos. s}{R}, \quad \frac{d^2u}{ds^2}=-\frac{\sin. s}{R^2}. \quad \frac{d^3u}{ds^3}=-\frac{\cos. s}{R^3} \text{ etc.}$$

whence

$$u=\sin. s=s-\frac{s^3}{2.3.R^3}+\frac{s^5}{2.3.4.5.R^5}-\text{ etc.}$$

(50) We may in a similar way develop an arc in terms of its sine and cosine.

Let s be an arc whose sine is u, then (Art. 47)

$$\frac{ds}{du}=\frac{1}{\sqrt{1-u^2}}, \quad \frac{d^2s}{du^2}=u(1-u^2)^{-\frac{3}{2}}$$

$$\frac{d^3s}{du^3}=(1-u^2)^{-\frac{3}{2}}+3u^2(1-u^2)^{-\frac{5}{2}}$$

making $u=0$

$$A=(s)=0, \quad B=\left(\frac{ds}{du}\right)=1, \quad C=\frac{1}{2}\left(\frac{d^2s}{du^2}\right)=0, \quad D=\frac{1}{2.3}\frac{d^3s}{du^3}=\frac{1}{2.3}$$

etc., hence

$$s=\text{arc whose sine is } u=u+\frac{u^3}{2.3}+\frac{3u^5}{2.4.5}+\text{ etc.}$$

If we make u equal to the sine of $30^\circ=\frac{1}{2}$ we have for the value of the arc

$$s=\tfrac{1}{2}+\frac{1}{2.3.2^3}+\frac{3}{2.4.5.2^5}+\frac{3.5}{2.4.6.7.2^7}+\text{ etc.}$$

the sum of which is 0.52359 nearly; and multiplying this by 6 we have the length of the arc of a semi-circle, thus

$$180^\circ=\pi=3.14154 \text{ nearly}$$

which is also the approximate ratio of the diameter of a circle to its circumference.

SECTION VII.

OF TANGENT AND NORMAL LINES TO ALGEBRAIC CURVES.

(51) We have seen (Art. 34) that when x and y represent the abscissa and ordinate of the curve, $\frac{dy}{dx}$ will represent the tangent of the angle made by the tangent line of the curve with the axis of abscissas. Now the equation of a line drawn through any given point is

$$y-y'=a(x-x')$$

in which y' and x' are the coordinates of the given point, and a the tangent of the angle made by the line with the axis of abscissas. Hence for any curve in which x' and y' are the coordinates of the point of tangency, the equation of the tangent line through that point will be

$$y-y'=\frac{dy'}{dx'}(x-x')$$

The value of $\frac{dy'}{dx'}$ will, of course, be obtained from the equation of the curve, and by substituting that value we obtain the equation for the tangent line of that curve.

EXAMPLES.

Ex. 1. From the equation of the circle we have

$$\frac{dy'}{dx'}=-\frac{x'}{y'}$$

and hence

$$y-y'=-\frac{x'}{y'}(x-x')$$

or

$$yy'+xx'=R^2$$

becomes the equation of the tangent line to a circle.

Ex. 2. In the case of the parabola we have

$$\frac{dy'}{dx'}=\frac{p}{y'}$$

whence

$$y-y'=\frac{p}{y'}(x-x')$$

or

$$yy'=p(x+x')$$

becomes the equation of the line tangent to a parabola.

Ex. 3. The equation of the ellipse gives

$$\frac{dy'}{dx'}=-\frac{B^2x'}{A^2y'}$$

whence

$$y-y'=-\frac{B^2x'}{A^2y'}(x-x')$$

or

$$A^2yy'+B^2xx'=A^2B^2$$

becomes the equation of the line tangent to the ellipse.

Ex. 4. From the equation of the hyperbola referred to its center and asymptotes, we have

$$\frac{dy'}{dx'}=-\frac{y'}{x'}$$

whence

$$y-y'=-\frac{y'}{x'}(x-x')$$

or

$$yx'+xy'=\frac{A^2+B^2}{2}$$

becomes the equation of the line tangent to the hyperbola, referred to its center and asymptotes as coordinate axes — as in Art. 35.

Since the normal line is perpendicular to the tangent, if a' represent the tangent of *its* angle with the axis of abscissas, then it will be equal to $-\frac{1}{a}$ where a represents the tangent of the angle of inclination of the *tangent* line. Hence

$$a' = -\frac{dx'}{dy'}$$

and substituting this value in the equation

$$y - y' = a'(x - x')$$

we have

$$y - y' = -\frac{dx'}{dy'}(x - x')$$

for the general equation for the normal line, and it may be found for any particular curve by obtaining the value of $-\frac{dx'}{dy'}$ from the equation of the curve, and making the substitution as in the case of a tangent line.

Proposition I.

(52) To find the general expression for the length of the subtangent to any curve.

Let AP (Fig. 23) be any curve of which PT is the tangent at the point P, TB the subtangent, PN the normal, and PB the ordinate; then from the triangle TPB we have

$$PB = TB \,.\, \text{tang. } PTB$$

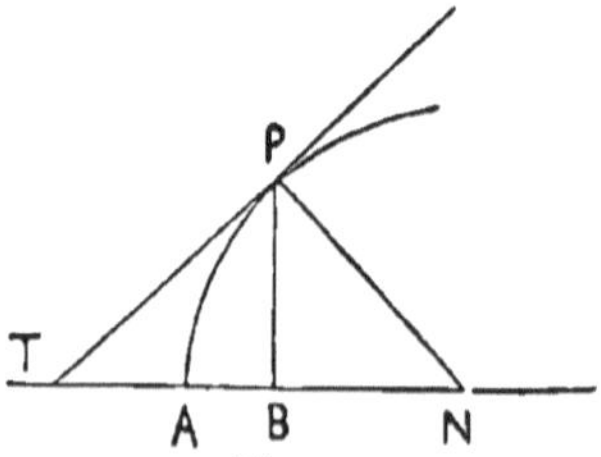

whence

$$TB = \frac{PB}{\text{tang. } PTB}$$

Fig. 23.

but

$$\text{tang. } PTB = \frac{dy}{dx} \text{ and } PB = y$$

hence

$$TB=y\frac{dx}{dy}$$

that is,

The subtangent to any curve is equal to the ordinate into the differential of the abscissa divided by the differential of the ordinate.

PROPOSITION II.

(53) To find the general expression for the length of the tangent to a curve.

From the triangle PTB (Fig. 23) we have

$$PT=\sqrt{\overline{PB}^2+\overline{TB}^2}$$

whence

$$PT=\sqrt{y^2+\frac{y^2\,dx^2}{dy^2}}=y\sqrt{1+\frac{dx^2}{dy^2}}$$

that is,

The length of the tangent to any curve is equal to the ordinate into the square root of one plus the square of the differential coefficient of the abscissa.

NOTE.—By the "*length of the tangent*" is meant that part of the tangent line between the point where it intersects the axis of abscissas and the point of tangency on the curve.

PROPOSITION III.

(54) To find the length of the subnormal to any curve.

Since the triangle PBN (Fig. 23) is similar to the triangle PBT, we have the angle BPN=BTP, and hence

$$BN=PB\,.\,\text{tang. }BPN$$

or

$$BN=y\frac{dy}{dx}$$

that is,

The subnormal is equal to the ordinate into the differential coefficient of the ordinate.

PROPOSITION IV.

(55) To find the length of the normal to any curve.

Since $\overline{PN}^2$ (Fig. 23) is equal to $\overline{PB}^2+\overline{BN}^2$, we have

$$PN=\sqrt{y^2+y^2\frac{dy^2}{dx^2}}=y\sqrt{1+\frac{dy^2}{dx^2}}$$

that is,

The length of the normal line is equal to the ordinate into the square root of one plus the square of the differential coefficient of the ordinate.

NOTE.—By the "*length of the normal*" is meant that part of it which lies between the point of its intersection with the axis of abscissas and the point of the curve to which it is drawn.

(56) The following examples will show the application of these formulas to particular cases.

Ex. 1. From the equation of the circle we have

$$\frac{dx}{dy}=-\frac{y}{x}$$

hence the subtangent (Fig. 24) is

$$TB=y\frac{dx}{dy}=-\frac{y^2}{x}=\frac{\overline{PB}^2}{BO}$$

a result that we also obtain from geometry.

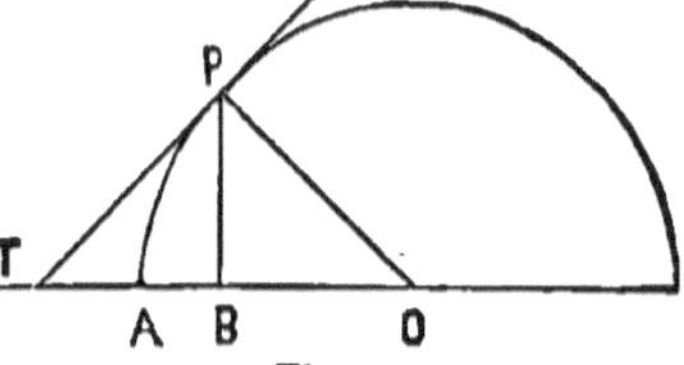

Fig. 24.

Ex. 2. The length of the tangent to the circle is

$$TP=y\sqrt{1+\frac{dx^2}{dy^2}}=y\sqrt{1+\frac{y^2}{x^2}}=\frac{y}{x}\sqrt{x^2+y^2}=\frac{Ry}{x}$$

We have also by geometry

$$TP:PO::PB:BO$$

whence

$$TP = \frac{PO \cdot PB}{BO} = \frac{Ry}{x}$$

Ex. 3. The normal line of the circle is

$$PO = y\sqrt{1 + \frac{dy^2}{dx^2}} = y\sqrt{1 + \frac{x^2}{y^2}} = R$$

Ex. 4. The subnormal of the circle is

$$BO = y\frac{dy}{dx} = -y\frac{x}{y} = -x$$

Ex. 5. From the equation of the parabola we have

$$\frac{dx}{dy} = \frac{y}{p}$$

hence the subtangent (Fig. 25) is

$$TB = y\frac{dx}{dy} = \frac{y^2}{p} = \frac{2px}{p} = 2x = 2AB$$

a result which we have also from geometry.

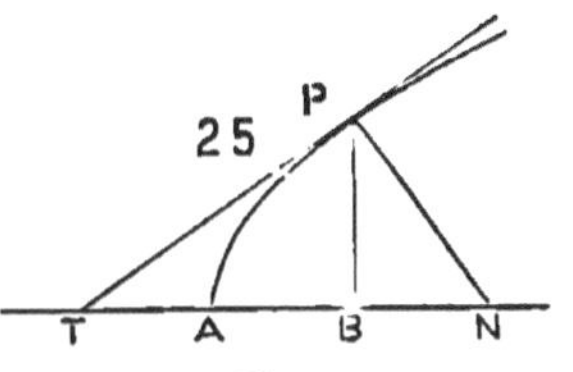

Fig. 25.

Ex. 6. The tangent of the parabola is

$$TP = y\sqrt{1 + \frac{dx^2}{dy^2}} = y\sqrt{1 + \frac{y^2}{p^2}} = \sqrt{y^2 + \frac{y^4}{p^2}} = \sqrt{y^2 + 4x^2}$$

We have just seen (Ex. 5) that $TB = 2AB$, hence

$$\overline{TB}^2 = 4\overline{AB}^2 = 4x^2$$

whence

$$TP = \sqrt{y^2 + 4x^2} = \sqrt{\overline{PB}^2 + \overline{TB}^2}$$

as is evident from the figure.

Ex. 7. The subnormal to the parabola is

$$BN = y\frac{dy}{dx} = y\frac{p}{y} = p$$

as we find from geometry.

Ex. 8. The normal to the parabola is

$$PN = y\sqrt{1+\frac{dy^2}{dx^2}} = y\sqrt{1+\frac{p^2}{y^2}} = \sqrt{y^2+p^2} = \sqrt{\overline{PB}^2+\overline{BN}^2}$$

which is evident from the figure.

Ex. 9. From the equation of the ellipse we have

$$\frac{dy}{dx} = -\frac{B^2x}{A^2y}$$

and the subtangent (Fig. 26) is

$$BT = y\frac{dx}{dy} = \frac{A^2y^2}{B^2x} = \frac{A^2-x^2}{x}$$

This value for the subtangent does not contain B, and hence is the same for all ellipses having the same major axis, the abscissa being the same. Hence the tangent to the circle at P will intersect the axis of abscissas at T, and

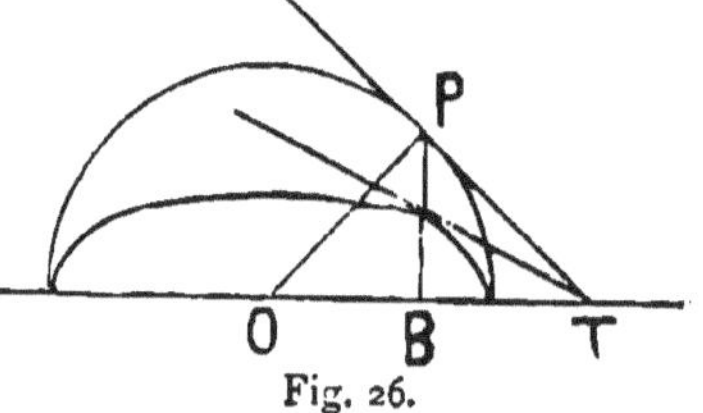

Fig. 26.

$$BT = \frac{\overline{PB}^2}{OB} = \frac{\overline{OP}^2 - \overline{OB}^2}{OB} = \frac{A^2-x^2}{x}$$

as we have already found.

SECTION VIII.

DIFFERENTIALS OF CURVES.

(57) We have seen (Art. 34) that the differential of a curve is equal to the square root of the sum of the squares of the differentials of the ordinate and abscissa. Hence to find the differential of any particular curve, we must find from its equation, the differential of one of the coordinates in terms of the other. The formula will then give the differential of the curve in terms of a single variable.

EXAMPLES.

Ex. 1. From the equation of the circle we have

$$dy=-\frac{xdx}{y}=-\frac{xdx}{\sqrt{R^2-x^2}}$$

hence, if we designate the arc by u, we have

$$du=\sqrt{dx^2+dy^2}=\sqrt{dx^2+\frac{x^2dx^2}{R^2-x^2}}=\frac{Rdx}{\sqrt{R^2-x^2}}$$

which is the differential of the arc of a circle in terms of the variable abscissa.

Ex. 2. From the equation of the parabola we have

$$dx=\frac{y}{p}dy$$

and calling the length of the arc u we have

$$du=\sqrt{dx^2+dy^2}=\sqrt{dy^2+\frac{y^2dy^2}{p^2}}=\frac{dy}{p}\sqrt{p^2+y^2}$$

Ex. 3. From the equation of the ellipse we have

$$y^2 = \frac{B^2}{A^2}(A^2 - x^2) \text{ and } dy = -\frac{B^2 x}{A^2 y}dx$$

hence

$$dy^2 = \frac{B^4 x^2}{A^4 y^2}dx^2 = \frac{B^4 x^2 dx^2}{A^4 \frac{B^2}{A^2}(A^2 - x^2)} = \frac{B^2 x^2 dx^2}{A^2(A^2 - x^2)}$$

hence

$$du = \sqrt{dx^2 + dy^2} = \frac{1}{A}dx\sqrt{\frac{A^4 - (A^2 - B^2)x^2}{A^2 - x^2}}$$

DIFFERENTIALS OF PLANE SURFACES.

(58) Every surface may be considered as generated by the flowing of a line.

If we wish to obtain the *rate* at which the surface is generated we must, if possible, consider every point in the line to be moving in a direction perpendicular to the line itself, if it is straight, or to its tangent at that point if it is a curve. For the only method of estimating the rate at which the surface is generated is by means of the length of the generating line and the rate with which it moves. Now unless the movement is made in a direction perpendicular to the line, the rate of *its* motion will be no criterion of the rate with which the surface is generated. Thus the line AB (Fig. 27) moving in a direction perpendicular to itself will generate the rectangle ABCD, but if it move at the same rate in any other direction as A*d*, the surface generated in the same time will be less until if it should move in its own direction it would generate no surface whatever. Hence in order that the *simple* movement of the line may be properly an element in estimating the rate of generation of the surface, it must always be supposed to take place in a

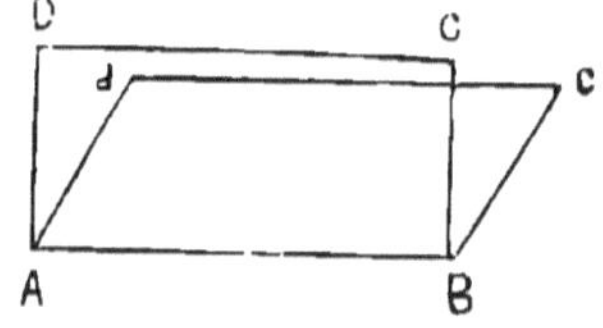

Fig. 27.

direction perpendicular to the line itself at every point. Otherwise we must include in our estimate of the rate, the sine of the angle made by the line with the direction in which it moves, which in most cases would be inconvenient, and, in many, impracticable.

(59) A plane surface may be generated in two ways by a straight line — by moving so as to be always parallel to itself, or by revolving about a fixed point. If it is supposed to be generated by the first method, and the boundary line is symmetrical about the axis of abscissas, the ordinate of the line is taken as the generatrix, and while it moves parallel to itself one of the extremities is in the line, and the other in the axis, and thus half the surface is generated.

Thus if we consider AB, the diameter of the circle ADB (Fig. 28) as the axis of abscissas, we would consider the upper half of the circle as generated by the ordinate DE moving parallel to itself, one extremity being always in the curve and the other in the axis AB. And similarly with a surface bounded by any other line that is symmetrical about the axis of abscissas.

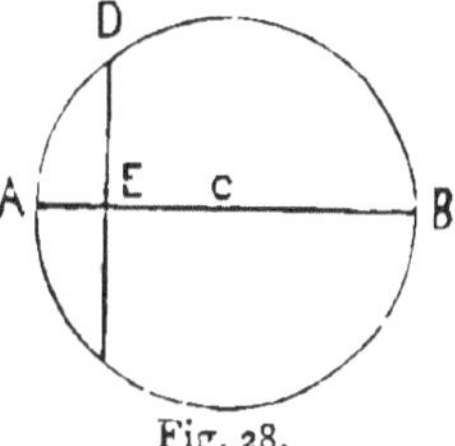

Fig. 28.

(60) The differential or rate of increase of any surface at the moment the generating line has arrived at any given position, such as BC (Fig. 29), will be represented by the increment that *would take place* (Art. 2) in a unit of time if the surface should increase *uniformly*, after the generating line should leave the position BC at the same rate as that with which it arrived there. Now, in order that the increment may be uniform, the generating line must maintain the same length and flow at an unvarying rate. Thus let AC be the curve and CB the generating line of the surface ACB; and let B*b* represent the

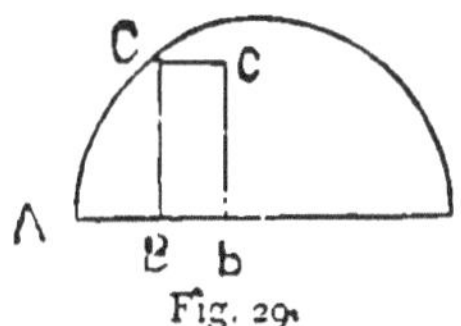

Fig. 29.

uniform increment of AB in a unit of time, at the same rate as at B; then the rectangle C*cb*B will represent what *would be* the uniform increment of the surface during the same unit of time at the rate at which it was increasing at CB. And, hence, if we consider the increment B*b* as the symbol representing the rate of increase of AB, the rectangle C*cb*B will be the proper symbol to represent the rate of increase or differential of the surface ACB. But the rectangle is equal to BC . B*b*; and if we call AB x, BC will be y, and B*b* the differential of x. Hence C*cb*B, or the differential of the surface will be

$$ydx$$

that is

The differential of a plane surface bounded by the axis of abscissas and a curve, is equal to the ordinate multiplied by the differential of the abscissa.

(61) In order to obtain the differential of any particular plane surface we must know the equation of the line that bounds it, in order that we may eliminate x or y from the formula. We shall then have the differential of the surface in terms of a single variable.

Example 1.

(62) To find the differential of a triangle.

Let ABC (Fig. 30) be the triangle, referred to A as the origin and AB and AD as coordinate axes. The equation of the line AC is

$$y=ax$$

hence

$$ydx=axdx$$

which is the differential of the surface of the triangle; a being the tangent of the angle made by the line AC with the axis AB.

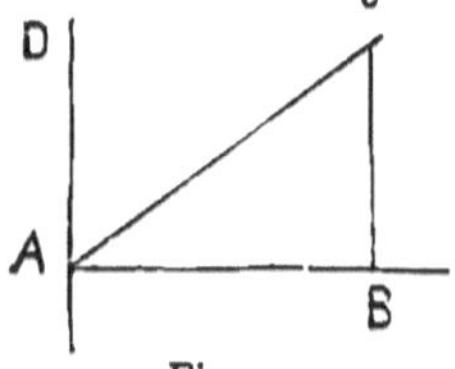

Fig. 30.

EXAMPLE 2.

(63) To find the differential of the surface of a semi-circle.

If we take the equation of the circle with the origin at the extremity of the diameter, we have

$$y=\sqrt{2Rx-x^2}$$

whence

$$ydx=dx\sqrt{2Rx-x^2}$$

which is the differential of the surface of a semicircle, the origin being at the extremity of the diameter. If we take the origin at the center we have

$$ydx=dx\sqrt{R^2-x^2}$$

EXAMPLE 3.

(64) To find the differential of the surface of a semi-ellipse.

If the ellipse be referred to its center and axes, we have from its equation

$$y=\frac{B}{A}\sqrt{A^2-x^2}$$

hence

$$ydx=\frac{B}{A}dx\sqrt{A^2-x^2}$$

which is the differential of the surface of the semi-ellipse.

EXAMPLE 4.

(65) To find the differential of the surface of a semi-parabola.

From the equation of the parabola referred to its vertex and axis we have

$$y=\sqrt{2px}$$

hence

$$ydx=dx\sqrt{2px}=\sqrt{2p}\,x^{\frac{1}{2}}dx$$

which is the differential of the surface of a semi-parabola.

DIFFERENTIALS OF SURFACES OF REVOLUTION.

(66) A surface of revolution is one which may be generated by a curve revolving about a line in the same plane. Every point in the revolving curve will describe a circle whose plane is perpendicular to the axis of revolution and whose center is in the axis. Any plane passed through the axis will cut from the surface a curve which is identical with the revolving curve.

Such a surface may also be supposed to be generated by the circumference of a circular section, made by a plane passed through the surface perpendicular to the axis, moving parallel to itself with its center in the axis of revolution, and its radius varying in such a manner, that its circumference shall always intersect the meridian section or directing curve.

The rate of increase, or differential, will be determined, as in other cases, by finding the surface that *would be generated* in a unit of time, if the generating circle were to move during that time, without change of magnitude at a uniform rate, equal to that with which it arrived at the point of differentiation. Such a surface would be equal to the circumference of the generating circle into the line which represents its rate of motion. Now the center of the generating circle is supposed to move along the axis at a uniform rate, hence its circumference will move along the directing curve at the same rate as the generating point of the curve; so that the line which represents this rate will be the same as the differential of the curve.

Moreover the suppositive differential surface that we are

seeking must be generated at a *uniform rate*, and hence the diameter of the generating circle must not change; so that the surface will be that of a cylinder, whose base is the circumference of the generating circle at the point of differentiation, and its height, the line which represents the differential of the directing curve at the same point.

If now we take the axis of abscissas as the axis of revolution, the radius of the generating circle will be an ordinate of the directing curve and the differential of the curve will be $\sqrt{dx^2+dy^2}$ (Art. 34); and hence calling the surface of revolution S, we have

$$dS=2\pi y\sqrt{dx^2+dy^2}$$

that is

The differential of a surface of revolution is equal to the circumference of the generating circle into the differential of the directing curve.

To apply this formula we obtain from the equation of the directing curve, the value of one variable in terms of the other, and by substitution obtain the differential in terms of a single independent variable.

Example 1.

(67) To find the differential of the surface of a cone.

In this case the revolving line is straight, and not a curve, but the principles of the rule apply equally well.

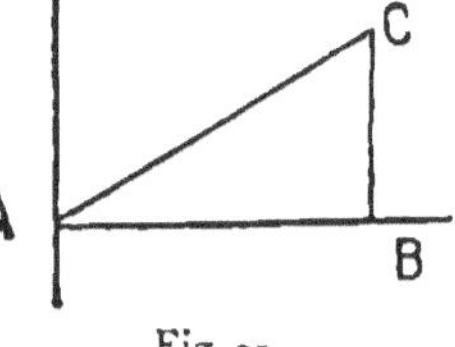

Fig. 31.

Let AC (Fig. 31) be the revolving element of the cone, and AB the axis of revolution and of abscissas, the origin being at A. Then we have for the equation of the line AC, $y=ax$ and $dy=adx$, a being the tangent of the angle BAC.

Substituting these values in the formula we have

$$dS=2\pi ax\sqrt{a^2dx^2+dx^2}=2\pi axdx\sqrt{a^2+1}$$

EXAMPLE 2.

(68) To find the differential of the surface of a sphere.

From the equation of the circle we have

$$dy=-\frac{xdx}{y} \text{ and } dy^2=\frac{x^2dx^2}{y^2}$$

hence

$$2\pi y\sqrt{dx^2+dy^2}=2\pi y\sqrt{\frac{x^2+y^2}{y^2}dx^2}$$

whence

$$dS=2\pi Rdx$$

for the differential of the surface of a sphere.

As the entire expression besides dx is composed of constants, we infer that the surface of a sphere increases at the same rate as the axis.

EXAMPLE 3.

(69) To find the differential of the surface of a paraboloid of revolution.

From the equation of the parabola we have

$$dx=\frac{ydy}{p} \text{ and } dx^2=\frac{y^2dy^2}{p^2}$$

hence

$$dS=2\pi y\sqrt{dx^2+dy^2}=2\pi y\sqrt{\frac{y^2+p^2}{p^2}dy^2}$$

EXAMPLE 4.

(70) To find the differential of the surface of an ellipsoid of revolution.

We found (Art. 57) that the differential of the elliptic curve is

$$\frac{1}{A}dx\sqrt{\frac{A^4-(A^2-B^2)x^2}{A^2-x^2}}$$

hence if we substitute this expression in place of $\sqrt{dx^2+dy^2}$

in the formula, and for y its value derived from the equation of the ellipse, we have

$$dS=2\pi\sqrt{\frac{B^2}{A^2}(A^2-x^2)}\cdot\frac{1}{A}dx\sqrt{\frac{A^4-(A^2-B^2)x^2}{A^2-x^2}}$$

which becomes by reduction

$$dS=\frac{2\pi B dx}{A^2}\sqrt{A^4-(A^2-B^2)x^2}$$

DIFFERENTIALS OF SOLIDS OF REVOLUTION.

(71) A solid of revolution is one which is described or generated by a surface, bounded by a line and the axis about which it revolves. If this axis be that of abscissas, then the ordinates of the bounding line will describe circles, of which they will be the radii and the centers will be in the axis. Any one of these circles may be considered as the generatrix, which describes the solid by moving parallel to itself, as in the last case. But it is now the surface of the circle and not merely its circumference that generates; and its movement is measured along the axis, the rate being the same as that by which the abscissa of the directing curve is increasing.

Now the rate of increase of a solid of revolution is measured by a suppositive increment that *would be described* in a unit of time, by the generating circle moving uniformly along the axis, with its diameter unchanged at the same rate as that with which the abscissa is generated. Hence such a solid would be a cylinder whose base is the generating circle, and whose altitude is the line representing the differential of the abscissa. But the area of the generating circle is πy^2, and the altitude of the cylinder is dx; hence the cylinder representing the differential of a solid of revolution, would be expressed by the function,

$$\pi y^2 dx$$

hence,

The differential of a solid of revolution is equal to the generating circle multiplied by the differential of the abscissa of the bounding line.

EXAMPLE 1.

(72) To find the differential of the volume of a cone.

If we take the vertex of the cone for the origin, and the axis of abscissas for its axis, the equation of the revolving line will be

$$y=ax$$

and hence calling v the volume of the cone we have

$$dv=\pi y^2 dx=\pi a^2 x^2 dx$$

in which a is the tangent of the angle made by the revolving line with the axis, and x the distance from the vertex to the base of the cone.

EXAMPLE 2.

(73) To find the differential of the volume of a sphere.

If we take the origin at the extremity of the diameter, the equation of the revolving semi-circle will be

$$y^2=2Rx-x^2$$

in which R is the radius of the sphere, and x any portion of the axis of revolution measured from its extremity at the origin until it equals 2R; hence the formula for the differential becomes

$$dv=\pi y^2 dx=\pi(2Rx-x^2)dx$$

EXAMPLE 3.

(74) To find the differential of the volume of an ellipsoid of revolution.

If we suppose the semi-ellipse to revolve about its major

axis, it will generate an oblong ellipsoid of revolution, otherwise called a prolate spheroid. If we take the origin at the extremity of the transverse axis, the equation of the ellipse is

$$y^2 = \frac{B^2}{A^2}(2Ax - x^2)$$

and hence the formula for the differential of the volume becomes

$$dv = \pi y^2 dx = \pi \frac{B^2}{A^2}(2Ax - x^2)dx$$

in which A is the semi-transverse and B the semi-conjugate axis of the ellipse which generates the ellipsoid of revolution.

If we take the conjugate axis of the ellipse for the axis of revolution and its extremity for the origin, we have

$$y^2 = \frac{A^2}{B^2}(2Bx - x^2)$$

and

$$dv = \pi \frac{A^2}{B^2}(2Bx - x^2)dx$$

In this case the volume is an oblate ellipsoid, or otherwise, an oblate spheroid.

Example 5.

(75) To find the differential of the volume of a paraboloid of revolution.

The axis of the parabola being the axis of revolution, and the origin at the vertex, we have

$$dv = \pi y^2 dx = 2\pi p x dx$$

in which p is the semi-parameter of the revolving parabola that generates the volume.

SECTION IX.

POLAR CURVES.

PROPOSITION I.

(76) To find the tangent of the angle which the tangent line makes with the radius vector.

Let CC′ (Fig. 32) be any curve of which we have the polar equation. Let P be the pole; PM $=r$ the radius vector; P$b=$R, the radius of the measuring arc bd; bPd, the variable angle $=v$; and OT, the tangent to the curve at the point M. Produce the radius vector, PM to R, draw RO perpendicular to PR, meeting the tangent in O; draw ON parallel to RM, and MN parallel to RO, meeting each other in N. Join PN, and draw ab parallel to MN meeting PN in a. Suppose the radius vector to revolve around the point P in the direction from d towards b.

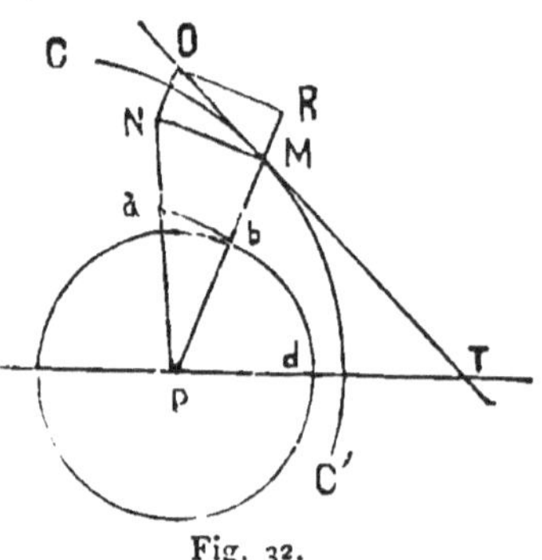

Fig. 32.

The generating point, being supposed to have arrived at the point M of the curve will be subject to two distinct, although mutually dependent, laws or tendencies. One of these tendencies arises from the law of change in the length of the radius vector, which causes the generating point to move outward in the direction of its length. The other tendency arises from the revolving motion of the radius vector which causes every point in it (including, of course, the generating point) to move in a direction perpendicular to itself. Hence in this case, the law would incline the generating point to move in the direction MN. If then we take the distance MR to represent the uniform outward movement that *would take place*, under the influence of the first law in a unit of time, it will represent the rate of change in the radius vector arising from that law, and is, therefore, the symbol of that rate; that is

$$MR = dr$$

If we take MN to represent what *would be* the uniform movement under the second law in the same length of time, it will represent the rate with which it tends to move in the direction MN arising from *that* law. Now as both these laws act together without disturbing each other, the generating point, if left to its tendency at the point M would move in such a direction as to obey both laws or influences at the same time; and hence at the end of the same unit of time would be found at O, having described the line MO; the departure from the line MN being ON=MR, and the departure from the line MR being RO=MN. But the generating point of a curve, if left to its tendency at any time would move in a line tangent to the curve, and since the line MO would be uniformly described in a unit of time, it represents the rate of increase of the curve, and is also tangent to it. Hence if we call the length of the curve u we have

$$MO = du$$

The point b at the intersection of the radius vector with the arc of the measuring circle, *tends to move* in the direction ba, and if left to that tendency would describe that line in the same time that the generating point would describe the line MN; for the rate of movement of b is to that of M as Pb is to PM, or as ba is to MN. If then we consider b as a point in the arc of the measuring circle, we may consider ba as representing its rate of increase, that is the rate of increase of the angle bPd, and hence

$$ab = dv$$

But from the triangles PMN and Pba we have

$$Pb : PM :: ab : MN$$

hence

$$MN = \frac{PM \,.\, ab}{Pb} = \frac{rdv}{R}$$

Now the tangent of the angle PMT = MON is equal to

$$R\frac{MN}{NO} = R\frac{MN}{MR}$$

and substituting the value of MN just found and of MR, we have

$$\text{Tang. } PMT = \frac{rdv}{dr}$$

that is

The tangent of the angle which the line tangent to a polar curve makes with the radius vector is equal to the radius vector into the differential of the measuring angle divided by that of the radius vector.

(77) Since MNO (Fig. 32) is a right angled triangle, we have

$$\overline{MO}^2 = \overline{MN}^2 + \overline{NO}^2 = \overline{MN}^2 + \overline{MR}^2$$

hence by substitution

$$du^2 = \frac{r^2 dv^2}{R^2} + dr^2$$

whence

$$du = \frac{1}{R}\sqrt{r^2 dv^2 + R^2 dr^2}$$

or making R = 1

$$du = \sqrt{r^2 dv^2 + dr^2}$$

that is

The differential of the arc of a polar curve is equal to the square root of the sum of the squares of the radius vector into the differential of the measuring angle, and of the differential of the radius vector.

PROPOSITION II.

(78) To find the subtangent of a polar curve.

The subtangent of a polar curve is the projection of the tangent on a line drawn through the pole perpendicular to the radius vector of the point of tangency.

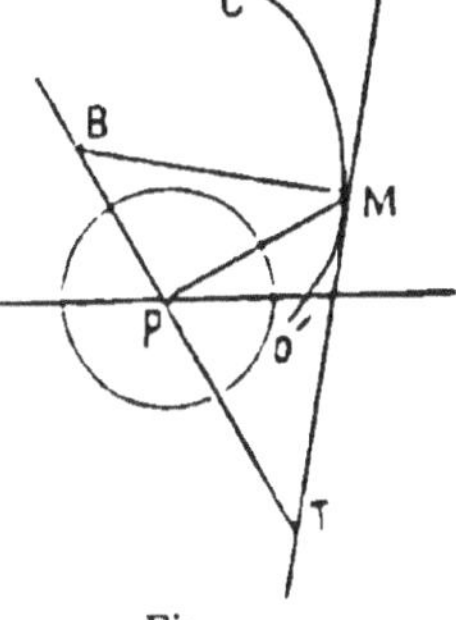

Fig. 33.

Hence if PT (Fig. 33) be drawn perpendicular to PM, meeting in T the tangent to the curve at the point M, then PT will be the subtangent. Since

$$PT = \frac{MP \,.\, \text{tang. } PMT}{R}$$

we have by substitution

$$PT = \frac{r}{R} \cdot \frac{rdv}{dr} = \frac{r^2 dv}{Rdr}$$

that is

The subtangent of a polar curve is equal to the square of the radius vector into the differential of the measuring arc divided by R into the differential of the radius vector. If we make $R = r$ we have $PT = \dfrac{rdv}{dr}$ = tangent of PMT.

PROPOSITION III.

(79) To find the value of the tangent to a polar curve

The tangent to a polar curve is that part of the tangent line which lies between its intersection with the subtangent and the point of tangency.

Hence MT (Fig. 33) will represent the tangent, and since

$$\overline{MT}^2 = \overline{PM}^2 + \overline{PT}^2$$

we have

$$\overline{MT}^2 = r^2 + \frac{r^4 dv^2}{R^2 dr^2}$$

or

$$MT = r\sqrt{1 + \frac{r^2 dv^2}{R^2 dr^2}} = \frac{r}{R dr}\sqrt{R^2 dr^2 + r^2 dv^2}$$

or making $R = 1$

$$MT = r\sqrt{1 + \frac{r^2 dv^2}{dr^2}}$$

PROPOSITION IV.

(80) To find the subnormal to a polar curve.

The subnormal of a polar curve is the projection of the normal line on a line drawn through the pole perpendicular to the radius vector for that point of the curve to which the normal is drawn.

Hence if MB (Fig. 33) be a normal at the point M, BP will be the subnormal.

The triangles MBP and MTP being similar, the angles MBP and PMT are equal, and since

$$PM = BP \frac{\text{tang. } MBP}{R}$$

we have

$$r = BP \frac{r dv}{R dr}$$

or

$$BP = \frac{R dr}{dv}$$

that is

The subnormal of a polar curve is equal to radius into the differential of the radius vector, divided by the differential of the measuring arc.

(81) The normal line MB (Fig. 33) is equal to

$$\sqrt{\overline{MP}^2 + \overline{PB}^2}$$

or

$$MB=\sqrt{r^2+\frac{R^2dr^2}{dv^2}}$$

(82) While the point at the extremity of the radius vector describes the line of a polar curve, the radius vector itself generates the surface bounded by the curve.

Now the point M of the line PM (Fig. 32) tends, by virtue of the revolving motion of the radius vector about the pole, to move in the direction MN, perpendicular to PM, and every other point in the line PM will tend to move in a direction parallel to MN, and at a rate proportional to its distance from the fixed point P. Hence if the point M were to be found at N, the line PM would assume the position PN, and the triangle PMN would be that which would be generated at a uniform rate by the radius vector PM if left to its tendency when in that position, so that the triangle PMN is the true symbol to represent the rate at which the surface bounded by the polar curve is generated, or, designating the surface by O, we have

$$\text{triangle } PMN=dO.$$

But

$$PMN=\frac{PM\times MN}{2}.$$

and substituting here the values already found for these terms we have

$$dO=\frac{r^2dv}{2R}$$

hence

The differential of a surface bounded by a polar curve is equal to the square of the radius vector into the differential of the measuring arc divided by twice its radius.

SPIRALS.

(83) If a right line revolve uniformly in the same plane about one of its points, and a second point should, at the same time approach to, or recede from the fixed point, according to some prescribed law, it would generate a curve called a spiral.

The fixed point is called the *pole*, and the curve generated during one revolution of the line is called a *spire*. There being no limit to the number of revolutions of the line, the number of spires is infinite, and a line, drawn through the pole will intersect the curve in an infinite number of points.

Hence there can be no algebraic relation between the ordinates and abscissas of the curve, and its conditions must be expressed by a polar equation which will be in the form

$$r = F(v)$$

in which r is the radius vector and v the measuring arc of the variable angle.

SPIRAL OF ARCHIMEDES

(84) This spiral is one in which the radius vector is constantly proportional to the corresponding arc which measures its angular movement. Hence its equation will be

$$r = av \qquad (1)$$

The curve may be constructed in the following manner. Divide the circumference of the measuring circle into eight equal parts by the radii AB, AC, AD, AE, etc., (Fig. 34); also the radius AB into the same number of parts. Then lay off from the center one of these parts on AC, two on AD, three on AE, and so on, there being eight on AB, nine on AC, ten on AD, and so on.

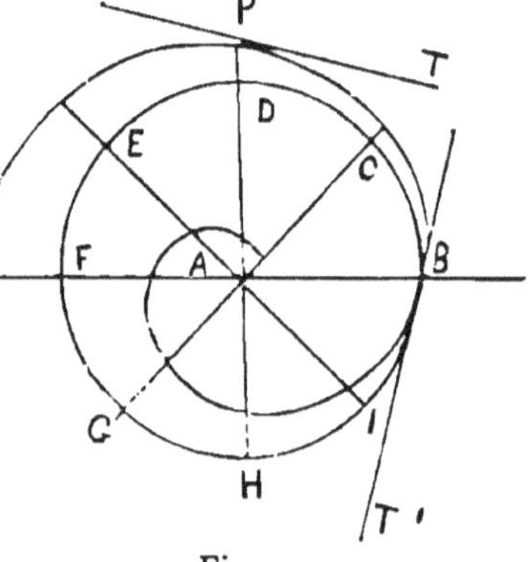

Fig. 34.

Through the points thus found draw the curve commencing

at the pole. The radius vector will be to the corresponding measuring arc as the radius of the measuring circle is to the circumference; or a will be equal to $\frac{1}{2\pi}$.

NOTE.— In this construction we have supposed the radius of the measuring circle to be equal to the radius vector after one revolution. Of course any other proportion might be taken, but as the magnitude of the spiral does not depend on that of the measuring circle, the radius of the latter may always be taken equal to the radius vector after one revolution.

If we differentiate equation (1) we have

$$dr = a\,dv \qquad (2)$$

In a polar curve (Art. 76) the tangent of the angle which the tangent line makes with the radius vector is equal to

$$r\frac{dv}{dr}$$

and from equation (2) we have

$$\frac{dv}{dr} = \frac{1}{a}$$

hence the tangent of the angle APT is equal to $\frac{r}{a}$, or in this case, to

$$2\pi r$$

This tangent will after one revolution be equal to

$$2\pi R$$

(85) The subtangent of a polar curve (Art. 78) is

$$\frac{r^2 dv}{R\,dr}$$

which becomes for this curve

$$\frac{r^2}{Ra} = \frac{2\pi r^2}{R}$$

or, making $r = R$, we have

$$\text{subtangent} = 2\pi R$$

equal to the tangent of the angle made by the tangent line with the radius vector; and also equal to the circumference of the circle described by the radius vector as a radius, when the point of tangency is at the circumference of the measuring circle.

If we make $v=n2\pi R$, that is, if the tangent be drawn to the curve after n revolutions of the radius vector, then

$$a=\frac{r}{v}=\frac{r}{n2\pi R}$$

whence

$$\frac{dv}{dr}=\frac{1}{a}=\frac{n2\pi R}{r} \text{ and } \frac{r^2 dv}{R dr}=n2\pi r$$

that is

After n revolutions of the radius vector, the subtangent is equal to n times the circumference of a circle described by the radius vector as a radius.

For the subnormal whose value is (Art. 80)

$$\frac{R dr}{dv}$$

we have

$$\frac{dr}{dv}=a=\frac{r}{v}$$

hence

$$\text{subnormal}=\frac{Rr}{v}$$

If the normal is drawn at the point B then

$$v=2\pi R$$

and we have

$$\text{subnormal}=\frac{r}{2\pi}$$

that is

The subnormal is equal to the radius of a circle of which $r=R$ is the circumference.

THE HYPERBOLIC SPIRAL.

(86) The equation of the Hyperbolic Spiral is

$$rv=ab$$

in which r is the radius vector, v the measuring arc, a the radius of the measuring circle and b the unit of the measuring arc — ab being, of course, a constant quantity. It is called a *Hyperbolic* spiral because its equation resembles that of a hyperbola referred to its center and asymptotes.

To construct this curve describe a circle with a radius PA (Fig. 35) equal to a. Lay off from A an arc $AB = b$ as the unit of the measuring arc v, and continue this division around the circumference of the measuring circle. Also lay off $As = \frac{1}{2}b$, $Ar = \frac{1}{4}b$, and so on.

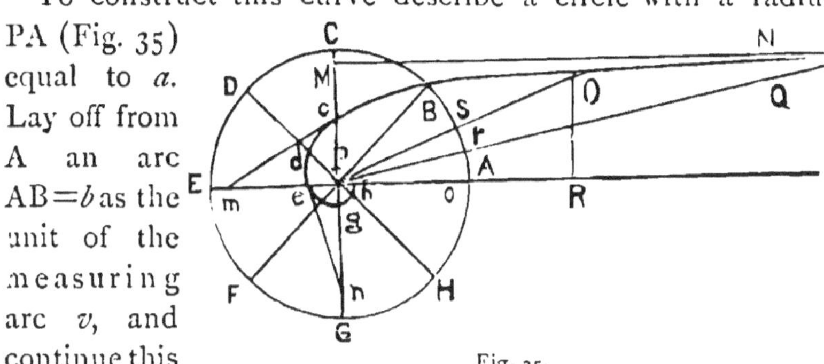

Fig. 35.

Through these points of division in the circumference draw the radii PB, PC, PD, PE, and so on, and produce the radii Ps and Pr. On these radii lay off $Pc = \frac{1}{2}a$, $Pd = \frac{1}{3}a$, $Pe = \frac{1}{4}a$, and so on; also $PO = 2a$, $PQ = 4a$. Draw the curve through the points thus found.

The radius vector multiplied by the measuring arc, counting from A, will always be equal to the radius of the measuring circle into the unit of the arc, that is

$$rv = ab$$

We see from the equation that r increases as v diminishes, and *vice versa*. If $v = o$ r becomes infinite, and hence the radius vector, through A, will never reach the beginning of the curve. If $r = o$ then v will be infinite, hence the curve will never reach the pole.

If we take any point O in the spiral and join OP, then OP will be equal to r, and the arc $As = v$. Draw OR perpendicular to PA and we have

$$OR=\frac{OP\,.\,\sin.\,As}{PA}=\frac{r\sin.\,v}{a}$$

hence

$$r=\frac{OR\,.\,a}{\sin.\,v}$$

and since

$$rv=ab$$

we have

$$\frac{OR\,.\,av}{\sin.\,v}=ab$$

whence

$$OR=\frac{\sin.\,v}{v}b$$

As sin. v is always less than v, the line OR will always be less than b, but may be made to approach that value as near as we please. Hence if we draw a line MN parallel to PA, at a distance from it equal to b=arc AB, it will be an asymptote of the curve.

Since

$$\frac{dv}{dr}=-\frac{v}{r}$$

the subtangent (Art. 78)

$$\frac{r^2dv}{Rdr}=-\frac{rv}{a}$$

and since

$$rv=ab$$

we have

$$\text{subtangent}=-b=-\text{arc AB}$$

a constant quantity. Thus Pm or Pn=AB or PM

Also the tangent of the angle made by the tangent line with the radius vector (Art. 76).

$$\frac{rdv}{dr}=-v$$

that is,

The tangent of the angle which the tangent line makes with the radius vector is negative and equal to the arc which measures the angle made by the radius vector with the fixed line PA. Hence this angle is obtuse on the side of the radius vector toward the origin, while the subtangent, being also negative, lies on the side *opposite* to the origin.

THE LOGARITHMIC SPIRAL.

(87) The equation of the logarithmic spiral is

$$v = \text{Log. } r$$

in which v represents the measuring arc and r the radius vector.

The equation may also be put into the form

$$a^v = r$$

the relation between v and r being such that while v increases in *arithmetical* progression r will increase in *geometrical* progression. Hence the curve may be constructed by laying off on the measuring arc the equal distances Ab, bc, cd, de, and so on (Fig. 36), and through the points of division drawing the radii Pb, Pc, Pd, Pe, and so on, producing them if necessary. On these radii lay off the distances PB, PC, PD, PE, and so on in geometrical progression, so that

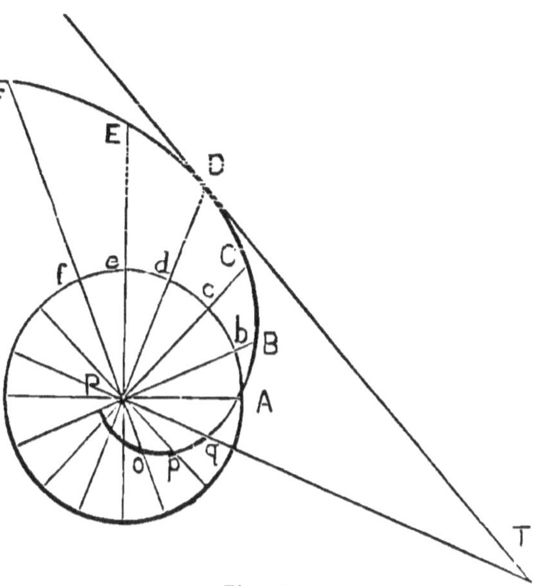

Fig. 36.

$$\frac{\text{PB}}{\text{PA}} = \frac{\text{PC}}{\text{PB}} = \frac{\text{PD}}{\text{PC}} = \frac{\text{PE}}{\text{PD}} \text{ and so on}$$

and through the points thus found draw the curve. That part of the curve within the circle will be found by laying off on the radii Pq, Pp, Po, and so on, distances from P by the same rule, and thus points of the curve may be found.

If we make the radius of the measuring circle equal to 1, and reckon the arc v from the line PA, then the curve will pass through the point A. for when $v=o$ we have

$$\log. r=o=\log. 1$$

and if we call a the ratio between PA and PB, we shall have

$$PB=a,\ PC=a^2,\ PD=a^3,\ PE=a^4, \text{ etc.}$$

when the exponent is always equal to the number of divisions of the measuring arc, and is therefore represented by the arc itself corresponding to the radius vector, whence

$$a^v=r \text{ or } v=\text{Log. } r \text{ to the base } a.$$

If we differentiate the equation of this curve we have

$$dv=M\frac{dr}{r} \qquad (2)$$

whence (Art. 76)

$$\text{tang. PDT}=\frac{rdv}{dr}=\frac{rMdr}{rdr}=M$$

that is

The tangent of the angle made by the tangent line with the radius vector is constant and equal to the modulus of the system of logarithms to which the curve belongs. If the system is the Naperian, M=1 and the angle PDT is equal to 45°.

The formula for the subtangent of a polar curve (Art. 78) is

$$\frac{r^2dv}{Rdr}$$

and substituting in this the value of $\frac{dv}{dr}$ from equation (2) we have (R being 1)

$$\text{subtan.}=\frac{r^2M}{r}=rM$$

If M=1 then subtang.$=r$.

For the value of the tangent we have (Art. 79)

$$\text{tang.}=\sqrt{r^2+r^2M^2}=r\sqrt{1+M^2}$$

If M=1 then

$$\text{tang}=r\sqrt{2}$$

For the subnormal we have (Art. 80)

$$\text{subnormal} = \frac{dr}{dv} = \frac{r}{M}$$

If $M=1$ then

$$\text{subnormal} = r$$

For the value of the normal (Art. 81) we have

$$\text{normal} = \sqrt{r^2 + \frac{r^2}{M^2}} = r\sqrt{1 + \frac{1}{M^2}}$$

If $M=1$ then

$$\text{normal} = r\sqrt{2}$$

These values show that these lines are all in direct proportion to the radius vector. The same result flows from the constancy of the angle made by the radius vector with the tangent line. For all the triangles formed by the radius vector, the tangent, and the subtangent will be similar to each other, at whatever point of the curve the tangent may be drawn. The same may be said of the triangles found by radius vector, normal and subnormal. Hence these lines will always be in proportion to the radius vector.

To construct a logarithmic spiral for a *given base*, describe a circle with a radius equal to a unit of the radius vector, PA, and lay off the arc A*b* equal to a unit of the measuring arc. Draw the radius vector PB equal to the given base; A and B will be points of the curve. Other points may be found as already described.

That part of the curve below the line PA corresponds to the negative value of v, and for that we have

$$r = \frac{1}{a^v}$$

in which when $r=o$, v will be infinite. Hence the curve is unlimited in both directions.

SECTION X.

ASYMPTOTES.

(88) An asymptote to a curve is a line, which the curve continually approaches, but never meets. Such a line is said to be tangent to the curve at an infinite distance, by which we are to understand that the point of contact to which the lines approach is beyond any finite limit.

That this may be the case it is necessary that, at least, one of the coordinates of the curve may have an unlimited value. Hence when we are seeking an asymptote to a curve, our first inquiry must be, whether the equation of the curve will admit of such values for the coordinates or either of them. If not, there can be no asymptotes. If it will do so for either coordinate, we must substitute that value in the equation and ascertain the resulting value for the other coordinate. If this resulting value is finite, there is an asymptote parallel to the axis of the infinite coordinate; if zero then the axis of the infinite coordinate is itself the asymptote. But if it should be infinite, then we must resort to the following method.

Find from the equation the values of the coordinates at the points where the tangent line intersects the axes, that is,

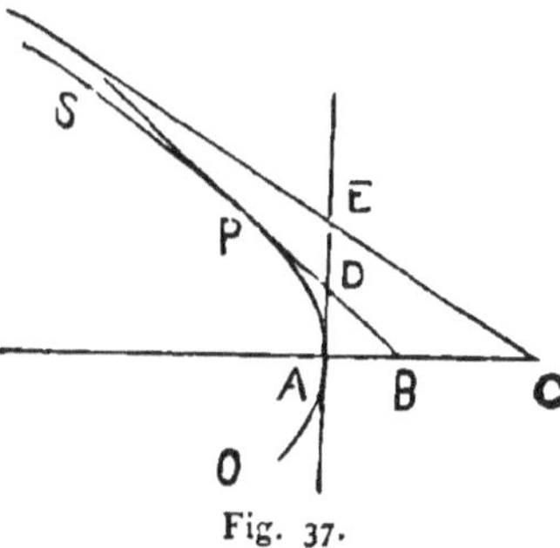

Fig. 37.

the distances from the origin. These points may be found as follows:

Let A (Fig. 37) be the origin of coordinates for the curve SO, and let PB be tangent to the curve at the point P, of which the coordinates are x' and y'. The equation of this tangent line is

$$y-y'=\frac{dy'}{dx'}(x-x')$$

If we make $y=o$ we have

$$x=x'-y'\frac{dx'}{dy'}=\text{AB}$$

If we make $x=o$ we have

$$y=y'-x'\frac{dy'}{dx'}=\text{AD}$$

If EC be an asymptote, and the values of x' and y' are made such as to remove the point of tangency to an infinite distance, then AB and AD will become AC and AE.

If in such case we have finite values for these distances, then there will be one or more asymptotes; if there is but one finite value, there will be one asymptote parallel to the axis of the infinite coordinate. If one be zero then the axis of the infinite coordinate is itself the asymptote. If both be zero then the asymptote passes through the origin; but if both be infinite there is no asymptote.

EXAMPLES.

Ex. 1. The equation of the hyperbola referred to its center and asymptotes is

$$xy=\text{M}$$

in which if x is made infinite y becomes zero; and if y is made infinite x becomes zero; hence both axes are asymptotes

Ex 2. If we consider the hyperbola as referred to its center and axes, its equation is

$$A^2y^2=B^2x^2-A^2B^2$$

where either x or y may be made infinite, and such value makes the other infinite also. Hence we take the formulas for the points of intersection of the tangent with the axes, which give

$$x=x'-y'\frac{A^2y'}{B^2x'}=-\frac{A^2y'^2-B^2x'^2}{B^2x'}=\frac{A^2}{x'}$$

and

$$y=y'-x'\frac{B^2x'}{A^2y'}=\frac{A^2y'^2-B^2x'^2}{A^2y'}=-\frac{B^2}{y'}$$

both of which values becomes zero, when x' and y' are made infinite. Hence the asymptotes pass through the origin.

Ex. 3. The equation of the parabola

$$y^2=2px$$

shows that x and y both become infinite together, and hence we take

$$x=x'-y'\frac{dx'}{dy'}=x'-\frac{y'^2}{p}=-x'$$

and

$$y=y'-x'\frac{dy'}{dx'}=y'-\frac{p}{y'}x'=\frac{y'}{2}$$

both of which values become infinite when x' and y' are infinite, and hence there is no asymptote to the parabola.

Ex. 4. If we take the ellipse whose equation is

$$A^2y^2+B^2x^2=A^2B^2$$

we see that neither x nor y can ever be infinite; in fact y can never exceed B nor x exceed A; hence there is no asymptote to the ellipse.

Ex. 5. The equation of the logarithmic curve is

$$x=\log.y$$

It may be constructed by laying off on the axis of abscissas (Fig. 38) the distances AB, AC, AD, etc., in arithmetical progression, and, on the corresponding ordinates, the distances Aa, Bb, Cc, Dd, etc., in geometrical progression, and drawing a curve through the points thus found. We see from the equation that if either x or y is infinite on the positive side, the other will be infinite also. If we apply the formula for the intersection of the tangent line with the axis we have (Art. 38)

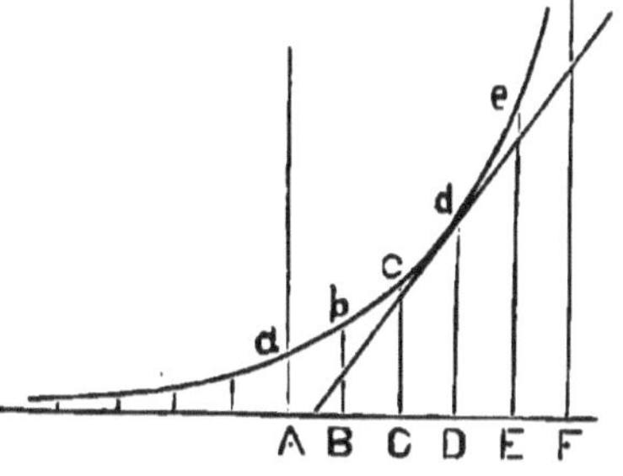

Fig. 38.

$$y=y'-x'\frac{dy'}{dx'}=y'-x'\frac{y'}{M}=y'\left(1-\frac{x'}{M}\right) \qquad (1)$$

and

$$x=x'-y'\frac{dx'}{dy'}=x'-y'\frac{M}{y'}=x'-M \qquad (2)$$

We see from these values, that when x' is infinite x will be infinite positively, and y negatively. Hence there is no asymptote on the positive side of x. But if x' be made infinite negatively, y' will become zero; for the logarithm of o is negative infinity, which shows that the axis of abscissas is an asymptote on the negative side. The value of y however in equation (1) becomes $-0\,\infty$, which is indefinite.

We learn from equation (2) that the tangent always intersects the axis of abscissas at a distance equal to M on the negative side of the ordinate of the point of tangency. Hence the subtangent is constant and equal to the modulus of the system to which the curve belongs. If $x'=M$, then x and y both become zero, and the tangent passes through the origin.

If we put the equation into the form

$$y=a^x$$

and make x negative it becomes

$$y=\frac{1}{a^x}$$

which makes $y=o$ when $x=\infty$; whence we infer that the axis of abscissas is an asymptote to the curve on the negative side, as already shown.

SECTION XI.

SIGNIFICATION OF THE SECOND DIFFERENTIAL COEFFICIENT.

SIGN OF THE SECOND DIFFERENTIAL COEFFICIENT.

(89) We have seen (Art. 36) that the first differential of the ordinate indicates by its *sign* whether the curve is leaving or approaching the axis of abscissas; and by its *value* it determines the *rate* of such approach or departure; that is, the tangent of the angle made by the tangent line with the axis of abscissas.

As the point of tangency moves along the curve, the rate of its approach to, or departure from, the axis of abscissas is constantly changing, and upon the rate of this change will depend the direction and amount of curvature of the curve.

Wherever the curve is situated with reference to the axis of abscissas, if its rate of departure is an increasing rate, or its rate of approach is a decreasing rate, then the curve is *convex* toward the axis of abscissas; while if its rate of departure is decreasing, or its rate of approach is increasing, it will be concave toward that axis.

(90) Now the second differential of the ordinate will determine by its sign whether the first is an increasing or decreasing function. If the latter is positive and increasing, or negative and decreasing, its rate of change (that is the second differential of the ordinate) will be positive (Art. 3); but if it is positive and decreasing, or negative and increasing, its rate of change is negative.

NOTE.— It will be remembered that the sign of the differential and that of its coefficient are always the same, since the differential of the independent variable is always uniform and positive.

(91) If, therefore, the second differential coefficient should be positive, the first must be either an increasing positive or a decreasing negative function (Art. 3). If the curve is on the positive side of the axis of abscissas, it is convex to that axis; if on the negative side it is concave.

(92) If the second differential coefficient is negative, the first must be either an increasing negative function, or a decreasing positive one. Hence the curve, if on the positive side of the axis of abscissas will be concave, and on the negative side convex to that axis.

(93) To illustrate these principles let us suppose the second differential coefficient to be positive, then the first must be a positive increasing or a negative decreasing function. The curves in Fig. 40 and 41 answer to these conditions, for from C to D the first differential coefficient is negative (Art. 36) and decreasing, while from D to E it is positive and increasing in both cases.

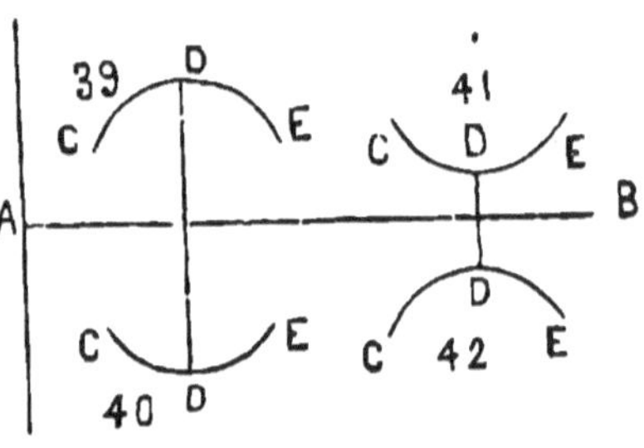

If the second differential coefficient is negative, then the first must be positive and decreasing, or negative and increasing, and we find the curves in Fig. 39 and 42 to answer these conditions; for from C to D the first differential coefficient (Art. 36) is positive and decreasing, while from D to E it is negative and increasing in both cases.

By inspecting these figures we see that for 39 and 40 the

second differential coefficient has in each case a sign contrary to that of the ordinate, and that both curves are concave to the axis AB; while in curves 41 and 42 the sign is the same as that of the ordinate, and the curves convex to the axis. Hence

When the signs of the second differential coefficient and of the ordinate are contrary to each other, the curve will be concave toward the axis of abscissas; when these signs are alike the curve will be convex toward that axis.

It will be noticed that in all these cases the first differential coefficient changes its sign at D where it becomes zero, but this does not affect the sign nor the value of the second differential, for the first may be changing as rapidly, and in either direction at the zero point as at any other.

(94) To illustrate these rules let us take the general equation of the circle

$$(x-a)^2+(y-b)^2=R^2$$

in which a is the abscissa and b the ordinate of the center.

Differentiating we have

$$\frac{dy}{dx}=-\frac{x-a}{y-b}$$

and

$$\frac{d^2y}{dx^2}=-\frac{R^2}{(y-b)^3}$$

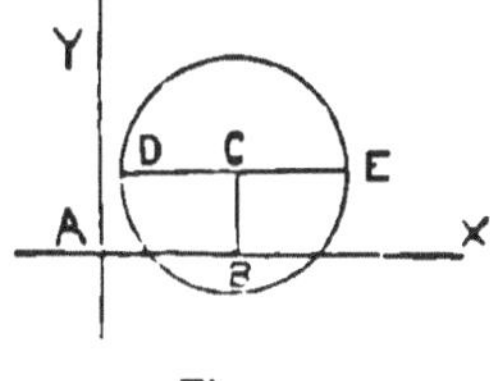

Fig. 43.

From which we learn that so long as y is greater than b the second differential coefficient will be negative, while it is positive where y is less than b, or where it is negative. We see also from the figure (Fig. 43) that above the line DE where y is greater than b the curve is concave toward the axis of abscissas, while between DE and the axis of abscissas, where y is positive and less than b, the curve is convex toward that axis. Below the axis of abscissas where y is negative the second differential is still positive, while the curve is concave toward the axis. All of which corresponds with the rule.

In the case of the parabola referred to its vertex and axis we have

$$\frac{d^2y}{dx^2} = -\frac{p^2}{y^3}$$

a fraction whose sign is always contrary to that of y; hence the curve is always concave towards the axis of abscissas.

The same may be said of the ellipse referred to its center and axes from whose equation we have

$$\frac{d^2y}{dx^2} = -\frac{B^4}{A^2y^3}$$

In the case of the hyperbola referred to its center and asymptotes we have

$$\frac{d^2y}{dx^2} = \frac{2y}{x^2}$$

a fraction whose sign is always the same as that of y. Hence the curve is everywhere convex toward the axis.

VALUE OF THE SECOND DIFFERENTIAL COEFFICIENT.

(95) The curvature of a curve at any point is the *tendency* of the tangent line at that point to change its direction, as the point of tangency is moving along the curve, in obedience to the *law of change* derived from the conditions which govern the movement of the generating point.

NOTE.— The curvature then of a curve is *not* "its deviation from the tangent,"* nor "its departure from the tangent drawn to the curve at that point,"† nor is it "the angular space between the curve and its tangent,"‡ nor is it any *actual* change in the direction of the tangent line as the point of tangency moves along the curve; nor does it depend on any such change, but upon the LAW which governs the movement of the generating point; for it is this law which fixes the *tendency* of the tangent to change its direction and this tendency is the curvature. Hence in estimating the curvature of a curve at any point, we consider that point *alone* and seek, *not* any actual movement of the generating point, but the *law which controls it.*

*Loomis. †Davies. ‡Church.

Hence if several curves as CD, C′D′, C″D″ (Fig. 44) have coincident tangents AB at the point A, and if we suppose the point of tangency to be at any instant moving along the curve, carrying with it its own tangent line, that one whose tangent line at the moment of coincidence is changing its direction most rapidly will have the greatest curvature at that point. For the *rate* of change in the direction of the tangent is the measure of its *tendency* to change.

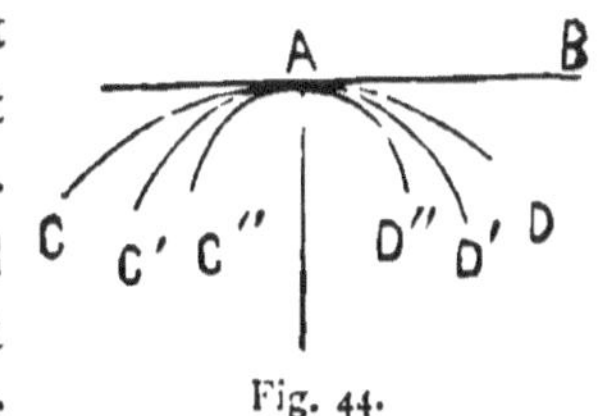

Fig. 44.

Since the first differential coefficient indicates the *direction* of the tangent to a curve, by means of the tangent of the angle made by it with the axis of abscissas; the second, which is simply the rate of change in the first, will indicate the rate at which the tangent of that angle is changing its value. Now as between two curves at a common tangent point, that curve in which the tangent line tends to change its *direction* most rapidly, will be the one in which the tangent of the angle made by that line with the axis of abscissas will also tend to change *its value* most rapidly, and will, therefore, have the greatest curvature, while if these tendencies are equal the curvatures are equal, and this will be indicated by the equality of the second differential coefficients.

SECTION XII.

CURVATURE OF LINES.

THEOREM.

(96) *The curvatures of different circles are inversely proportional to their radii.*

The curvature of a circle is the same at all points of the circumference, and all circles having the same radii have the same curvature.

Since the change in the direction of the tangent, as the point of tangency moves around the curve is constant, its *actual* change of direction for any given movement of the point of tangency, will always be in proportion to its *tendency* to change, multiplied by the length of the arc over which the movement is made, and may, therefore, be represented by that product; and hence the *tendency* to change or *curvature* will be equal to the actual change divided by the length of the arc.

Now the change in the direction of the tangent is equal to the angle contained between its two positions, which is the same as that contained between the two radii drawn to the extremities of the arc. Calling this angle v and the length of the arc a, we shall have

$$\text{curvature} = \frac{v}{a}$$

If now we have two circles, which we will call o and o', whose radii are r and r', and the angles at the center for the same length of arc a are v and v', we shall have

$$\text{curvature of } o = \frac{v}{a}$$

$$\text{curvature of } o' = \frac{v'}{a}$$

hence

$$\text{curvature of } o : \text{curvature of } o' :: v : v' \qquad (1)$$

but

$$v : 360^\circ :: a : 2\pi r$$

and

$$v' : 360^\circ :: a : 2\pi r'$$

whence

$$v \,.\, 2\pi r = v' \,.\, 2\pi r'$$

or

$$v : v' :: r' : r$$

Substituting this ratio in proportion (1) we have

$$\text{curvature of } o : \text{curvature of } o' :: r' : r$$

Q. E. D.

CONTACT OF CURVES.

(97) When two curves have a common point, the coordinates of that point must satisfy both their equations. This will generally be a point of *intersection*, and not a point of contact; and is all that can be secured by having but one condition common to the two curves.

If they are at the same time tangent to each other, at the common point, then another common condition is imposed and there is a contact of the first order.

The condition required in this case is, that, *for the point of contact*, the first differential coefficients shall be the same for the equations of both curves. For since the curves are tangent to each other, they have a common tangent line, and the first differential coefficient, which determines the angle made by this line with the axis of abscissas, must be the same for both equations.

If, besides this, the curves are required to have the same curvature at the point of contact, this will introduce a third condition, which is, that the *second* differential coefficients shall be the same for both equations (Art. 95).

For the second differential is the rate of change in the first, which gives the direction of the tangent line, and the rate of change in this direction is the curvature. This is a contact of the second order.

If now it is required, in addition, that the *rate of change in the curvature* should be the same in both curves at the point of contact; we must introduce a fourth condition, viz., that the *third* differential coefficient should be the same in both equations. This would be a contact of the third order. And thus the order of contact would become higher for every new condition introduced common to both curves, and every new agreement between the successive differential coefficients.

If then we wish to find the order of contact of two given curves, we first combine their equations, and determine their common point if they have one. For this point the variables will have the same value in both equations. If the values thus found being substituted in the first differential coefficient of each equation, reduce them to the same value, there is a contact of the first order; that is, they have a common tangent line at the common point.

If they also reduce the second differential coefficients of the two equations to the same value they have a contact of the second order, and so on for the successive differential coefficients; the order of contact being determined by the number of coefficients that successively become equal by the substitution of the values of the common coordinates.

EXAMPLE.

(98) To illustrate this rule let us take the two equations

$$4y=x^2-4 \qquad (1)$$

and

$$y^2-2y=3-x^2 \qquad (2)$$

from which we obtain by combination

$$y=-1 \text{ and } x=0$$

indicating that both the curves pass through the point of which these are the coordinates. We have also by differentiating twice — for equation (1)

$$\frac{dy}{dx}=\frac{x}{2} \text{ and } \frac{d^2y}{dx^2}=\tfrac{1}{2}$$

and for equation (2)

$$\frac{dy}{dx}=-\frac{x}{y-1} \text{ and } \frac{d^2y}{dx^2}=-\frac{1}{y-1}-\frac{x^2}{(y-1)^3}$$

Substituting in these differential coefficients the values of x and y just found, we have the first differential coefficients

$$\frac{dy}{dx}=\frac{x}{2}=0 \text{ and } \frac{dy}{dx}=-\frac{x}{y-1}=0$$

and the second differential coefficients

$$\frac{d^2y}{dx^2}=\tfrac{1}{2} \text{ and } \frac{d^2y}{dx^2}=-\frac{1}{y-1}-\frac{x^2}{(y-1)^3}=\tfrac{1}{2}$$

from which we infer that at the point whose coordinates are $x=0$ and $y=-1$, the curves have a contact of the second order. We also see from the value of the first differential coefficient that at that point the tangent to both curves is parallel to the axis of abscissas. A little investigation would show that the first curve is a parabola, and the second a circle tangent to the first at its vertex.

(99) The constants which enter into the equation of a curve determine the conditions which govern the movement of the generating point for that kind of curve; which must fulfil as many conditions as it has constants. Thus the circle whose general equation contains three constants, must fulfil three conditions, namely, two in the coordinates of the center, and one in the length of the radius. The ellipse must fulfil four conditions, namely, the coordinates of the center and the lengths of the two axes.

(100) Now if one curve be given complete by its equation with fixed values for its constants, and another with constants which are indeterminate, and capable of being adjusted to any given conditions, we may easily assign such values to them as will cause the curve to fulfil such conditions as may be required of it. We may, for instance, require the curve to pass through a given point in a given curve. This will require that the same variable coordinates shall satisfy the equations of both curves for that point. We may also require them to have a common tangent at that point; this will require the constants to be so adjusted that the first differential coefficients of the two equations shall be equal. If there are three or more constants in each equation we may require such values as will cause the second differential coefficients to become equal also, thus producing an equality of curvature, or a contact of the second order, at the common point. And thus we may continue until the order of contact is one less than the number of constants to be disposed of.

(101) In order to make this adaptation of the second curve to the first we must consider its constants, or as many of them as will be required for the purpose as unknown quantities (Art. 4) and construct as many equations as may be required to determine them.

These equations are derived from the conditions to be fulfilled by the constants. Thus the first which requires that the second curve shall pass through a point of the first will generally be met by the proper adjustment of a single constant; and an equation formed by substituting in that of the curve to be adjusted the values of the coordinates of the designated point, and also the values of the known constants, will determine the value of the unknown constant.

If it is required that the two curves be tangent to each other, we must adapt the values of *two* constants to this condition, and this is done by substituting the same values of the common coordinates, and of the remaining constants in the *first* differential coefficients of the two equations, and placing them equal to each other, thus forming a second equation. A contact of the second order may be secured by fixing the value of a third constant in a similar way by means of the *second* differential coefficients of the two equations.

The values of these constants thus determined being substituted in the general equation of the required curve, will produce an equation of one that will fulfill the required conditions; that is, one that will intersect at a given point, or have a contact of a required order.

EXAMPLE.

(102) To illustrate these principles let us take the equation of the ellipse referred to its center and axes

$$A^2y^2+B^2x^2=A^2B^2$$

and the general equation of the circle

$$(x-a)^2+(y-b)^2=R^2 \qquad (1)$$

in which the constants are arbitrary and may be adapted to any prescribed conditions. Suppose we say that the circumference shall pass through the upper extremity of the conjugate axis where

$$x=o \text{ and } y=B$$

This being but one condition will require the adaptation of but one constant. Let this be a, while we make $R=A$ and $b=o$.

Then substituting these values in equation (1) we have

$$(o-a)^2+(B-o)^2=A^2$$

or

$$a^2+B^2=A^2$$

whence

$$a=\pm\sqrt{A^2-B^2}$$

and the equation of the circle becomes

$$(x\mp\sqrt{A^2-B^2})^2+y^2=A^2$$

the center being in one of the foci — the plus value of the radical corresponding with the focus on the positive side of the center.

If we add another condition, namely, that the curves shall be tangent to each other at the same point, we must adapt the value of *two* constants to these two conditions. Let these constants be a and b, and make $R=2B$. Then we must construct an equation between the first differential coefficients of the curves; that is

$$\frac{B^2x}{A^2y}=\frac{x-a}{y-b} \qquad (2)$$

Substituting the values of x and y as before we have

$$\frac{B^2o}{A^2B}=\frac{o-a}{B-b}$$

hence

$$a=o$$

and substituting these values in equation (1), we have

$$(B-b)^2=4B^2$$

whence

$$b=-B$$

and the equation of the required circle becomes

$$x^2+(y+B)^2=4B^2$$

the center being at the lower extremity of the conjugate axis where $a=o$ and $b=-B$.

If now we add a still further condition there shall be a contact of the second order at the same point we must adapt the values of *three* constants to that condition, by forming a third equation, between the *second* differential coefficients, thus

$$\frac{B^4}{A^2\,y^3}=\frac{1+\frac{dy^2}{dx^2}}{y-b}=\frac{1+\frac{(x-a)^2}{(y-b)^2}}{y-b} \qquad (3)$$

Substituting, as before, the values of $x=o$ and $y=B$ in equations (1), (2), (3), we have three equations from which to determine the values of the three constants; thus

$$(o-a)^2+(B-b)^2=R^2$$

$$\frac{B^2 o}{A^2 B}=\frac{o-a}{B-b}$$

$$\frac{B^4}{A^2 B^3}=\frac{1+\frac{(o-a)^2}{(B-b)^2}}{B-b}$$

From the second we obtain

$$a=o$$

From the third we have

$$b=\frac{B^2-A^2}{B}$$

and substituting these values in the first we obtain

$$R=\frac{A^2}{B}$$

and the equation of the circle becomes

$$x^2+\left(y-\frac{B^2-A^2}{B}\right)^2=\frac{A^4}{B^2}$$

the radius being equal to half the parameter of the conjugate axis of the ellipse, and the center being in that axis prolonged in a negative direction.

(103) In this last case we have the highest order of contact of which the circle is capable, and hence the circle is called the osculatrix to the ellipse; or is said to be osculatory to it.

An osculatrix to a curve is one which has the highest order of contact with it, that any curve of the same kind as the osculatrix can have.

Since the number of constants limits the number of conditions that can be assigned to a curve, and since the passing of the curves through the same point is one condition, the order of contact can only be equal to the remaining number of possible conditions; namely, the number of constants, less one, which enter into the general equation; and this will be the same as the order of its highest differential.

EXAMPLES.

(**104**) *Ex.* 1. To find the equation of the circle osculatory to the parabola, whose equation is

$$y^2=4x \qquad (1)$$

at the point where the coordinates are

$$x=1 \text{ and } y=2.$$

Differentiating this equation we have

$$\frac{dy}{dx}=\frac{2}{y} \text{ and } \frac{d^2y}{dx^2}=-\frac{4}{y^3}$$

whence

$$\frac{2}{y}=-\frac{x-a}{y-b}$$

and

$$\frac{4}{y^3}=\frac{1+\left(\frac{x-a}{y-b}\right)^2}{y-b}$$

or

$$1=-\frac{1-a}{2-b} \qquad (2)$$

and

$$\tfrac{1}{2}=\frac{1+\left(\frac{1-a}{2-b}\right)^2}{2-b} \qquad (3)$$

Also from the general equation of the circle we have

$$(1-a)^2+(2-b)^2=R^2 \qquad (1)$$

and from these we find

$$R^2=32, \quad a=5, \quad b=-2$$

and the equation of the circle osculatory to the parabola at the given point is

$$(x-5)^2+(y+2)^2=32$$

Ex. 2. To find the circle osculatory to an equilateral hyperbola whose equation is

$$xy=8$$

at a point whose coordinates are

$$y=4 \text{ and } x=2.$$

By differentiating we have

$$\frac{dy}{dx}=-\frac{y}{x}=-2$$

and

$$\frac{d^2y}{dx^2}=\frac{2y}{x^2}=2$$

and from the general equation of the circle we have

$$(2-a)^2+(4-b)^2=R^2 \qquad (1)$$

$$\frac{2-a}{4-b}=2 \qquad (2)$$

$$\frac{1+\left(\frac{2-a}{4-b}\right)^2}{4-b}=-2 \qquad (3)$$

from which we obtain

$$R^2=\frac{125}{4} \quad a=7 \quad b=\frac{13}{2}$$

and the equation of the required circle will be

$$(x-7)^2+\left(y-\frac{13}{2}\right)^2=\frac{125}{4}$$

Ex. 3. Find the equation of the circle osculatory to the curve whose equation is

$$4y=x^2-4$$

at a point whose coordinates are

$$x=0 \quad y=-1$$

RADIUS OF CURVATURE.

(105) Since the curvature of a curve at any point is the same as that of its osculatory circle at that point, we call the radius of the osculatory circle the *radius of curvature* of the curve. And since the formulas for the equation of the osculatory circle may be applied to any point of a given curve, we may consider them as expressing the *general conditions* required of the osculatory circle.

These formulas, as we have seen, are

$$(x-a)^2+(y-b)^2=R^2 \qquad (1)$$

$$\frac{dy}{dx}=-\frac{x-a}{y-b} \qquad (2)$$

$$\frac{d^2y}{dx^2}=-\frac{1+\frac{dy^2}{dx^2}}{y-b} \qquad (3)$$

the two last may be written

$$x-a=-\frac{dy}{dx}(y-b) \qquad (2)$$

and

$$y-b=-\frac{dx^2+dy^2}{d^2y} \qquad (3)$$

If we represent the coordinates of any given point in a curve by x' and y', then for the osculatory circle we must have

$$x=x',\ y=y',\ \frac{dy}{dx}=\frac{dy'}{dx'},\ \frac{d^2y}{dx^2}=\frac{d^2y'}{dx'^2}$$

The quantities a and b represent the coordinates of the center of the osculatory circle, and R is its radius.

If we substitute in equation (2) the value of $y-b$, we have

$$x-a=\frac{dy}{dx}\left(\frac{dx^2+dy^2}{d^2y}\right)$$

whence equation (1) becomes

$$\frac{dy^2}{dx^2}\left(\frac{dx^2+dy^2}{d^2y}\right)^2+\left(\frac{dx^2+dy^2}{d^2y}\right)^2=R^2$$

from which we have

$$R=\pm\frac{(dx^2+dy^2)^{\frac{3}{2}}}{dxd^2y} \qquad (5)$$

which is the general expression for the value of the radius of curvature in terms of quantities belonging to a given curve.

If we denote the length of the curve by u we shall have

$$R=\frac{du^3}{dxd^2y}$$

(106) Since the curve and its osculatory circle have a common tangent, they will also have a common normal; and as the normal to the circle passes through the center, the normal to any curve at any point will pass through the center of the circle osculatory to it at that point.

This is also shown from equation (2) which is

$$\frac{dy}{dx}=-\frac{x-a}{y-b}$$

x and y being coordinates both to the given curve and to the osculatory circle at the point of contact, and a and b the coordinates of the center of the circle.

For since $\frac{dy}{dx}$ is the tangent of the angle made by the tangent line with the axis of abscissas, we shall have

$$-\frac{dx}{dy}=\frac{y-b}{x-a}$$

for the tangent of the angle made by the normal line with the same axis. But when a straight line passes through two points — x and y being the coordinates of one, and a and b the coordinates of the other — the tangent of the angle made by that line with the axis of abscissas will be $\frac{y-b}{x-a}$, and hence the normal to the curve, since it passes through the first point will also pass through the second — that is, the center of the osculatory circle.

And since from equation (3) we have

$$(y-b)\frac{d^2y}{dx^2}=-\left(1+\frac{dy^2}{dx^2}\right)$$

the value of the first member of the equation will be essentially negative, and hence we infer that $y-b$ and $\frac{d^2y}{dx^2}$ must have contrary signs. So that if $\frac{d^2y}{dx^2}$ is negative, b will be less than y, and, if positive, it will be greater. In the first case the curve will be concave toward the axis of abscissas, and b will be between the curve and that axis; while in the other case the curve will be convex toward the axis of abscissas, and b will be *beyond* it. Hence the center of the osculatory circle will be on the concave side of the curve.

(107) To find the general expression for the radius of curvature of the parabola, we differentiate its equation twice and obtain

$$ydy=pdx$$

and

$$yd^2y+dy^2=0$$

whence

$$dy=\frac{pdx}{y}$$

and

$$d^2y=-\frac{dy^2}{y}=-\frac{p^2dx^2}{y^3}$$

Substituting these values in the formula we have

$$R=\frac{\left(dx^2+\frac{p^2dx^2}{y^2}\right)^{\frac{3}{2}}}{-dx\frac{p^2dx^2}{y^3}}=\frac{[dx^2(y^2+p^2)]^{\frac{3}{2}}}{-p^2dx^3}=\frac{(2px+p^2)^{\frac{3}{2}}}{-p^2}$$

or, the cube of the normal (Art. 56) divided by the square of half the parameter.

If we make $x=0$ we have

$$R=-p$$

or half the parameter for the radius of curvature at the vertex. If we make $x=\frac{1}{2}p$ we have

$$R=-p\sqrt{8}$$

for the radius of curvature at the point where the ordinate through the focus meets the curve. As every other value of R is greater than that where $x=0$ it follows that the greatest curvature of the parabola is at the vertex.

(108) From the equation of the circle we have

$$dy=-\frac{xdx}{y}$$

and

$$d^2y=-\frac{R^2dx^2}{y^3}$$

and substituting these values in the formula we have

$$R=\frac{\left(dx^2+\frac{x^2dx^2}{y^2}\right)^{\frac{3}{2}}}{-\frac{R^2dx^3}{y^3}}=\frac{[(y^2+x^2)dx^2]^{\frac{3}{2}}}{-R^2dx^3}=-R$$

the radius of the circle as it should be.

(109) From the equation of the ellipse we have

$$dy=-\frac{B^2xdx}{A^2y}$$

and

$$d^2y=-\frac{A^2dy^2+B^2dx^2}{A^2y}$$

or substituting in the last equation the value of dy^2 we have

$$d^2y=-\frac{B^4dx^2}{A^2y^3}$$

These values being substituted in the formula

$$R=\frac{\left(dx^2+\frac{B^4x^2dx^2}{A^4y^2}\right)^{\frac{3}{2}}}{-\frac{B^4dx^3}{A^2y^3}}=\frac{\left(\frac{A^4y^2+B^4x^2}{A^4}\right)^{\frac{3}{2}}}{-\frac{B^4}{A^2}}$$

which is equal to the cube of the normal divided by the square of half the parameter as in the parabola.

If we make $x=A$ we have $y=0$ and

$$R=-\frac{B^2}{A}$$

If $y=B$ then $x=0$, and we have

$$R=-\frac{A^2}{B}$$

Hence the radius of curvature of the ellipse at the principal vertex is half the parameter of the transverse axis— that is the ordinate through the focus. At the vertex of the conjugate axis, the radius is half the parameter of that axis (Art. 102).

(110) The equation of the hyperbola referred to its center and asymptotes gives

$$dy=-\frac{ydx}{x}$$

and

$$d^2y=-\frac{2dxdy}{x}$$

Substituting these values in the formula we have after reducing.

$$R=\frac{(x^2+y^2)^{\frac{3}{2}}}{2xy}=\frac{2(x^2+y^2)^{\frac{3}{2}}}{A^2+B^2}$$

In the equilateral hyperbola, this value becomes equal to the cube of the radius vector divided by the square of the semi-axis.

SECTION XIII.

EVOLUTES.

(**111**) If we suppose a circle to roll along the concave side of a curve, being always tangent to it, and at the same time varying the length of its radius so as to be osculatory also, its center will describe a curve which is called the *evolute* of the given curve; and its variables will be the coordinates of that variable center. In other words, the evolute of any curve is the *locus* of the centers of all the circles that can be drawn osculatory to that curve.

The relation between the variables of the evolute can be determined and its equation found from the equation of the given curve, and the first and second differential coefficients derived from that equation; since these determine the position and length of the radius of curvature, and consequently the place of the center of the osculatory circle.

Since the coordinates of the point of tangency and the first and second differential coefficients are the same for the given curve and for the osculatory circle, we can at once determine two of the properties of the evolute.

(**112**) The first of these properties is, the radius of the osculatory circle is tangent to the evolute.

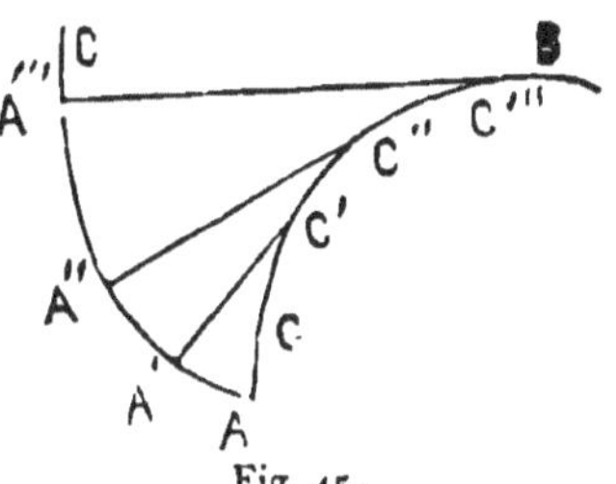

Fig. 45.

Let AC (Fig. 45) be any curve, and let c be the center of the osculatory circle for the point A, while c', c'', c''' are the centers of the osculatory circles corresponding to the points A′, A″, A‴. Then the curve cc''' passing through these centers will be the evolute, and any radius as $A''c''$ will be tangent to it at the point c'', the center of the osculatory circle.

The equations of conditions (Art. 105) may be put into the following form

$$(x-a)^2+(y-b)^2=R^2 \qquad (1)$$
$$(x-a)dx+(y-b)dy=0 \qquad (2)$$
$$(y-b)d^2y+dy^2+dx^2=0 \qquad (3)$$

and in this case a, b, R, x, y are variables; x being independent and dx a constant quantity; while x and y are coordinates of the given curve, and of the osculatory circle at the point of contact, and a and b coordinates of the variable center of the osculatory circle, that is, of the evolute, and are functions of x and y.

From these equations, as we have seen (Art. 105), R may be determined for any point in the given curve by eliminating a and b considered as constants. But for the evolute curve we must consider them as variable coordinates; and hence under that supposition if we differentiate equations (1) and (2) we have

$$(x-a)dx+(y-b)dy-(x-a)da-(y-b)db=RdR \qquad (4)$$

and

$$dx^2+dy^2+(y-b)d^2y-da\,.\,dx-db\,.\,dy=0 \qquad (5)$$

Subtracting equation (2) from (4) we have

$$-(x-a)da-(y-b)db=RdR \qquad (6)$$

and subtracting (3) from (5) we have

$$-da\,.\,dx-db\,.\,dy=0 \qquad (7)$$

whence

$$\frac{db}{da}=-\frac{dx}{dy} \qquad (8)$$

but $-\frac{dx}{dy}$ is the tangent of the angle made by the normal line to the curve, at the point whose coordinates are x and y, with the axis of abscissas; and $\frac{db}{da}$ is the tangent of the angle made by the tangent line to the curve, at the point whose coördinates are a and b, with the same axis. But x and y are coördinates of the given curve, which is now called the *involute*, and a and b are coördinates of the evolute, and, of course, of the center of the osculatory circle corresponding to the point $(x.y)$ on the curve, and through this center the normal line must pass (Art. 106); and since both the normal to the curve (or radius of curvature) and the tangent to the evolute pass through the same point, and make the same angle with the axis of abscissas, they must be one and the same line; and hence the proposition.

(113) The other property referred to in Art. 111 is

The difference between the length of the evolute curve and the radius of curvature of the involute, measured from the same point, is either zero or a constant quantity.

From equations (2) and (8), of the preceding article, we have

$$x-a=\frac{da}{db}(y-b) \tag{9}$$

and substituting this value of $x-a$ in equation (1) we have

$$(y-b)^2\frac{da^2}{db^2}+(y-b)^2=R^2=(y-b)^2\frac{da^2+db^2}{db^2} \tag{10}$$

From equations (9) and (6) we have

$$-\frac{da^2}{db}(y-b)-(y-b)db=RdR=-\frac{da^2+db^2}{db}(y-b)$$

which being squared gives

$$(y-b)^2\frac{(da^2+db^2)^2}{db^2}=R^2dR^2$$

and this being divided by equation (10) gives

$$da^2+db^2=dR^2$$

If we designate the length of the evolute by u we shall have

$$du^2=da^2+db^2$$

whence

$$du^2=dR^2$$

or

$$du=dR \text{ or } dR-du=o=d(R-u)$$

hence $R-u$ is a constant quantity and

$$R=u+c$$

If $u=o$ we have

$$R=c$$

and hence c is equal to the radius of curvature at the beginning of the curve, and R is at all times equal to the length of the evolute to the point where R is tangent plus the constant c.

If, therefore, we suppose a cora to be fastened at B (Fig. 45) and drawn tight around the curve AB and then unwound from A, the end of the cord will describe the curve AC of which the curve AB is the evolute. For the cord will be at all times tangent to the curve from which it is unwound, and also the momentary radius of the curve AC for the point at its own extremity, and consequently normal to the curve at that point; while the length of the cord from the point of tangency to its extremity in the curve AC is equal to the distance from the same point to the origin at A measured along the curve AB.

(114) To find the equation of the evolute, we must combine the equation of the osculatory circle with that of the involute in such a manner that $x \,.\, y$ and R shall disappear and leave an equation containing only a and b as variables.

This will require four equations, and these are obtained from the equation of the involute, the general equation of the circle, and those formed by placing the first and second differential coefficients of each of these equations respectively equal.

Thus if we take the equations of condition (Art. 105)

$$(x-a)^2+(y-b)^2=R^2 \qquad (1)$$

$$x-a=-\frac{dy}{dx}(y-b) \text{ or } \frac{dy}{dx}=-\frac{x-a}{y-b} \qquad (2)$$

$$y-b=-\frac{dx^2+dy^2}{d^2y} \quad \text{or} \quad \frac{d^2y}{dx^2}=-\frac{1+\frac{dy^2}{dx^2}}{y-b} \tag{3}$$

and then differentiating the equation of the involute twice, we find the values of the same differential coefficients and make them equal to the second members of equations (2) (3); then eliminate x, y and R, the resulting equation is that of the evolute.

Since R is contained in only one equation, we omit that, as the remaining three are sufficient for eliminating x and y, and for the resulting equation.

(115) To find the equation of the evolute to the parabola.

The equation of the parabola is

$$y^2=2px \tag{1}$$

from which

$$\frac{dy}{dx}=\frac{p}{y} \tag{2}$$

and

$$\frac{d^2y}{dx^2}=-\frac{p^2}{y^3} \tag{3}$$

Placing these differential coefficients equal to those of the general equation of the circle, we have

$$\frac{p}{y}=-\frac{x-a}{y-b} \tag{4}$$

and

$$\frac{p^2}{y^3}=\frac{1+\frac{dy^2}{dx^2}}{y-b} \tag{5}$$

Dividing equation (5) by equation (4), and substituting for y^2 its value from equation (1), and reducing, we have

$$a=3x+p \tag{6}$$

and substituting the values of a and y in equation (4) we have after reducing

$$b=-\frac{(2x)^{\frac{3}{2}}}{p^{\frac{1}{2}}}$$

and substituting in this the value of x from equation (6), and squaring, we have

$$b^2=\frac{2^3(a-p)^3}{3^3p}=\frac{8}{27p}(a-p)^3$$

which is the equation of the evolute of the parabola.

If we make $b=o$ we have $a=p$, which is the center of the osculatory circle for the vertex. If we transfer the origin to that point we have

$$a=p+a' \text{ and } b=b'$$

hence

$$b'^2=\frac{8}{27}a'^3$$

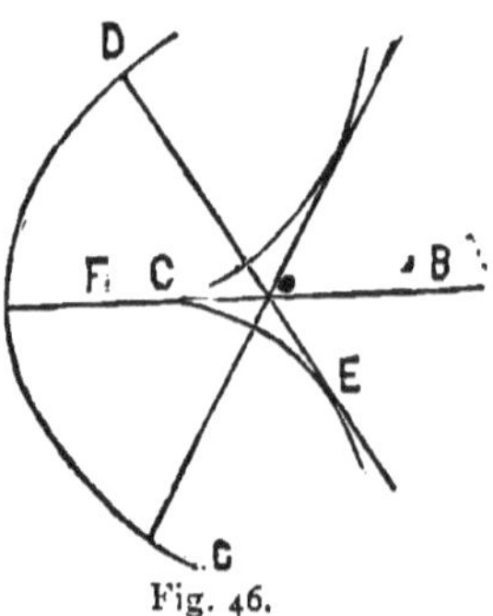

Fig. 46.

Since every value of a' gives two equal values for b' with contrary signs, the curve of the evolute ACE (Fig. 46) is symmetrical about the axis of abscissas. If a' is negative then b' is imaginary, and hence the curve commences at C, a point in the axis of abscissas at a distance from A equal to p—that is, at double the distance of the focus, or half the parameter.

(116) To find the equation of the evolute of the ellipse. For this case we have

$$A^2y^2+B^2x^2=A^2B^2 \tag{1}$$

$$\frac{dy}{dx}=-\frac{B^2x}{A^2y}=-\frac{x-a}{y-b} \tag{2}$$

$$\frac{d^2y}{dx^2}=-\frac{B^4}{A^2y^3}=-\frac{1+\frac{dy^2}{dx^2}}{y-b} \tag{3}$$

From equations (2) and (3) we have

$$y-b=\frac{A^2y(x-a)}{B^2x}=\frac{A^2y^3\left(1+\frac{dy^2}{dx^2}\right)}{B^4}$$

whence

$$\frac{x-a}{x}=\frac{y^2\left(1+\frac{dy^2}{dx^2}\right)}{B^2}=\frac{y^2\left(1+\frac{B^4x^2}{A^4y^2}\right)}{B^2}.$$

whence

$$B^2(x-a)=x\left(\frac{A^4y^2+B^4x^2}{A^4}\right)$$

whence

$$A^4B^2x-A^4B^2a=A^4xy^2+B^4x^3$$

whence

$$A^2x(A^2B^2-A^2y^2)=A^4B^2a+B^4x^3$$

whence

$$A^2B^2x^3=A^4B^2a+B^4x^3$$

whence

$$a=\frac{A^2-B^2}{A^4}x^3 \qquad (4)$$

Substituting this value of a in equation (2) we have

$$x-\frac{A^2-B^2}{A^4}x^3=\frac{(y-b)B^2x}{A^2y}$$

whence

$$\frac{A^4-(A^2-B^2)x^2}{A^2}=\frac{(y-b)B^2}{y}$$

whence

$$A^4y-A^2x^2y+B^2x^2y=A^2B^2y-A^2B^2b$$

whence

$$y(A^2-x^2-y^2)=-B^2b$$

Substituting for x^2 its value from equation (1) we have

$$y\left(A^2-\frac{A^2B^2-A^2y^2}{B^2}-y^2\right)=-B^2b$$

whence

$$b=-\frac{A^2-B^2}{B^4}y^3 \qquad (5)$$

Making $A^2-B^2=c^2$ we have

$$a=\frac{c^2}{A^4}x^3 \text{ and } b=-\frac{c^2}{B^4}y^3$$

and making $\frac{c^2}{A}=m$ and $\frac{c^2}{B}=n$, we have

$$\frac{a}{m}=\frac{x^3}{A^3} \text{ and } \frac{b}{n}=-\frac{y^3}{B^3}$$

or

$$\frac{x}{A}=\left(\frac{a}{m}\right)^{\frac{1}{3}} \text{ and } \frac{y}{B}=-\left(\frac{b}{n}\right)^{\frac{1}{3}}$$

Writing the equation of the ellipse under the form

$$\frac{x^2}{A^2}+\frac{y^2}{B^2}=1$$

and substituting the values of $\frac{x}{A}$ and $\frac{y}{B}$ just found, we have

$$\left(\frac{a}{m}\right)^{\frac{2}{3}}+\left(\frac{b}{n}\right)^{\frac{2}{3}}=1 \qquad (6)$$

which is the equation of the evolute of the ellipse in which a and b are the variable coordinates, and m and n the constants. If we make $a=0$ we have $b=\pm n$, and if $b=0$ we have $a=\pm m$, which shows that the form of the evolute is symmetrical with both axes of the ellipse. But

$$m=\frac{c^2}{A}=\frac{A^2-B^2}{A}$$

and subtracting this from A we have the radius of curvature at the principal vertex equal to $\frac{B^2}{A}$, as we have already seen (Art. 109). Similarly we find the radius of curvature at the vertex of the conjugate axis to be $\frac{A^2}{B}$.

If we differentiate equation (6) twice we have

$$\frac{1}{m}\left(\frac{a}{m}\right)^{-\frac{1}{3}}+\frac{1}{n}\left(\frac{b}{n}\right)^{-\frac{1}{3}}\frac{db}{da}=0$$

whence

$$\frac{db}{da}=-\frac{\frac{1}{m}\left(\frac{a}{m}\right)^{-\frac{1}{3}}}{\frac{1}{n}\left(\frac{b}{n}\right)^{-\frac{1}{3}}}=-\frac{n}{m}\left(\frac{an}{bm}\right)^{-\frac{1}{3}}$$

and

$$-\frac{1}{m^2}\left(\frac{a}{m}\right)^{-\frac{4}{3}}-\frac{1}{n^2}\left(\frac{b}{n}\right)^{-\frac{4}{3}}\frac{db^2}{da^2}+\frac{3}{n}\left(\frac{b}{n}\right)^{-\frac{1}{3}}\frac{d^2b}{da^2}=0$$

whence

$$\frac{d^2b}{da^2}=\frac{\frac{1}{m^2}\left(\frac{a}{m}\right)^{-\frac{4}{3}}+\frac{1}{n^2}\left(\frac{b}{n}\right)^{-\frac{4}{3}}\frac{db^2}{da^2}}{\frac{3}{n}\left(\frac{b}{n}\right)^{-\frac{1}{3}}}$$

Since the numerator of the second differential coefficient is always positive, it will have the same sign as the denominator, which is the same as that of b, and hence the curve is everywhere convex toward the axis of abscissas. The first differential coefficient becomes zero when $a=o$, and infinite when $b=o$, hence both axes are tangent to the curve, as in Fig. 47.

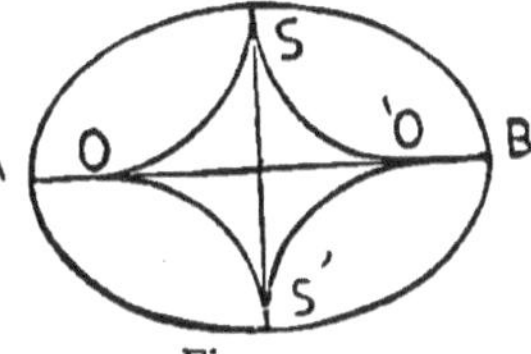

Fig. 47.

If we make A=B, then $c=o$, and also $m=o$ and $n=o$, hence a and b in equations (4) and (5) will also become zero as they should, since in case of the circle the evolute is reduced to a point — the center.

SECTION XIV.

ENVELOPES.

(117) Suppose two lines, AB and AC (Fig. 48), be drawn at right angles to each other, and a third line *ed* to move in such a manner that its extremities *d* and *e* shall be constantly in these axes, while its length remains unchanged; so that while the extremity *e* arrives successively at the points *e''*, *e'*, the extremity *d* will arrive at the corresponding points *d''*, *d'*.

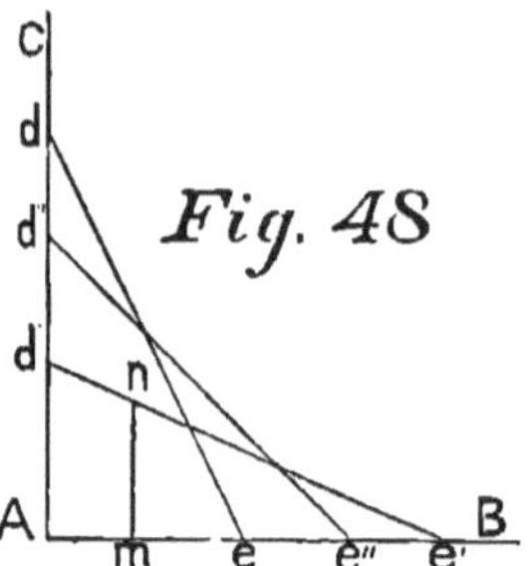

Fig. 48

During this movement those points of the line near the extremity *d* will move in the direction more nearly parallel to the axis AC than the line itself is, and will consequently fall *within* its first position, while the points near the extremity *e* will move in a direction more nearly parallel to the axis AB than the line is, and will consequently fall *without* its first position. But between these extreme points there is one that *tends to move in the direction of the line itself.*

This point does not, of course, remain fixed on the line, but moves from one extremity to the other as the line changes its position and direction, always occupying that place in the line which at the moment does not tend to move out of it towards either side. *The curve described by this point is the envelope of the curve.*

Again let AB (Fig. 49) be the transverse axis of an ellipse, and CD its conjugate axis; and suppose these axes to vary to any extent under the condition that the area of the ellipse shall remain constant.

Then as AB decreases CD will increase at a rate corresponding with this condition. When the curve thus commences to change its shape, a point near the extremity A will tend to move in a direction more nearly parallel to AB than the tangent to the curve at that point is, while a point near the extremity C will tend to move in a direction more nearly parallel with CD than the corresponding tangent line is. Now between A and C there is a point in the curve that tends (as the axes are changing) to move exactly in the direction of the tangent to the ellipse at that point.

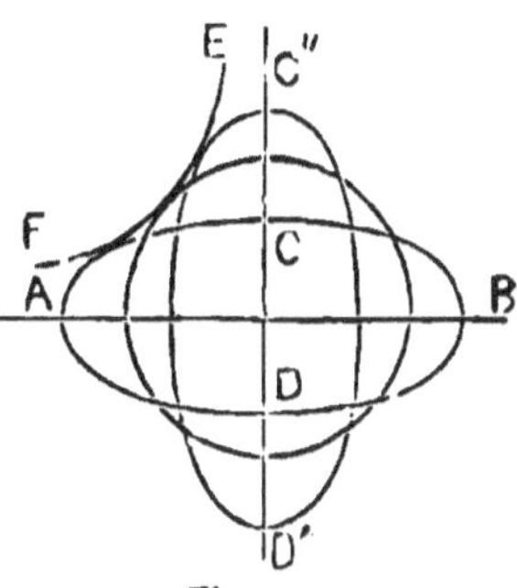

Fig. 49.

As the curve changes its shape and position this point will also change *its* place on the ellipse, keeping always where its *tendency* is in the direction of the tangent to the ellipse as it is at the moment. The movement of the point will be continuous, and it will generate a curve which will be the envelope of the ellipse.

(118) Since the point on the given curve which describes the envelope always tends to move in the direction of the momentary position of the tangent to the curve at that point, and since any generating point always tends to move in the direction of the tangent to its own curve, it follows that the given curve and its envelope will have a common tangent line wherever the generating point may be at the moment during the formation of the curve. Thus in the last illustration, the ellipse, in every stage of its change, will be tangent to the envelope at that point of the curve just then generated.

(119) *An envelope to any line, is another line generated by that point of the given line, which tends to move in the direction of the tangent, whenever its position or shape is made to change by changing the constants of its equation, or any of them, into variables.*

An envelope is not always produced by this change of the constants, for it may be that no point of the given line will tend to move in the direction of its tangent; as in the case of an ellipse where both axes are increased.

In general, there will be an envelope only where the successive positions of the line corresponding with minute changes in the constants, will *intersect each other;* for while the generating point of the envelope tends to move in the direction of the tangent, the points on each side of it will tend to move away from the tangent in opposite directions, hence the next position of the changing line will cross the previous one near the generating point of the envelope.

(120) If in any equation of a line the constants are made to vary in value, it is evident that while the curve or line remains the same in kind, its shape and position may assume every possible form and place within the limits determined by the law of variation imposed upon the constants of the equation.

If we take for example the ellipse, and consider A and B in its equation as independent variables, then

$$A^2y^2+B^2x^2=A^2B^2$$

will represent an infinite number of ellipses of every possible size and proportions subject to but two conditions; namely, the axes must both coincide with the axes of coordinates. If we make A and B dependent on each other we limit the system of ellipses by the condition thus introduced, but still their number is infinite. If we introduce the still further condition that the values of x and y shall be confined to those points of the system which tend to move in the direction of the tangent, while A and B tend to change their values, the first differential coefficient will not be affected by such tendency in A and B, and hence will be the same *at those points* whether they are considered as variables or constants. So then if we take the differential of the equation with respect to them only as variables, and make it equal to zero, and incorporate it with the original equation, we put this limit on the values of x and y, which will then only apply to points in the envelope. The equation will, therefore, be that of the envelope itself — that is, *instead of representing every point in one ellipse, it will represent one point in each quadrant of every ellipse that can be formed under the given conditions.*

To find the equation then of an envelope we differentiate the equation of the given line with reference to such only of the constants as are considered variable for the time being, and place that differential equal to zero. The values of the constants determined from this equation, and the conditions of relation among themselves, being substituted in the given equation, will produce one that will be independent of the variable constants, and this will be the equation of the envelope.

EXAMPLES.

(121) For the first example, let us take the general equation of the circle in which R and b are constants, while a is considered as a variable. Now since the values of x and y are to be confined to those points of the circle which tend to move in the direction of the tangent while a varies, it will make no difference whether we differentiate with respect to x and y only, or with respect to a also. Differentiating in both these ways we have

$$(x-a)dx+(y-b)dy=0$$

and

$$(x-a)dx-(x-a)da+(y-b)dy=0$$

making these differentials equal, and cancelling like terms we have

$$-(x-a)da=0 \tag{1}$$

which we should have obtained at once by differentiating with respect to a alone, considering all the rest as constants.

From equation (1) we have

$$x=a$$

and this value substituted in the general equation gives

$$y-b=\pm R \text{ or } y=b\pm R$$

If we take the positive value for R, this is the equation of a line DE (Fig. 50) parallel to the axis of abscissas at a distance equal to that of the centers of the system of circles plus the radius, and hence tangent to them all on the upper side, and is generated by the highest point of the circle as it moves from D to E, as a varies in value; that point being the one that tends to move (and in this case does move) in the direction of the tangent to the circle drawn through it. If we take the negative value of R, the equation represents the line D′E′ tangent to the system of circles on the lower side.

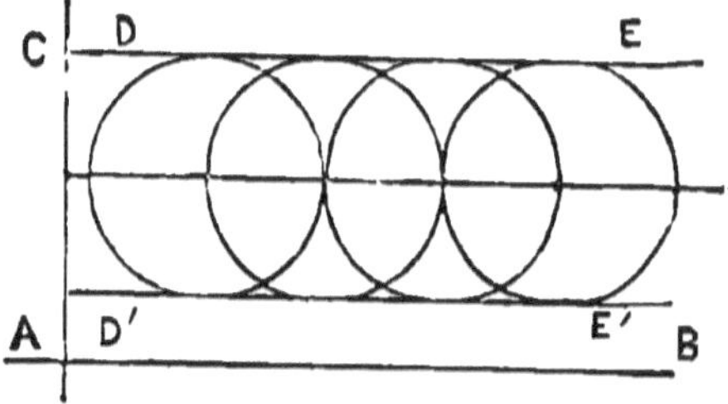

Fig. 50.

(122) If we take the same equation and consider a and b both as variables, we must establish a relation between them in order to make them both functions of x and y. Let this relation be expressed by the equation

$$a^2+b^2=c^2 \tag{1}$$

then the two equations will represent a system of circles

(Fig. 51) whose centers lie in the circumference of another circle whose radius is equal to c, and its center is at the origin.

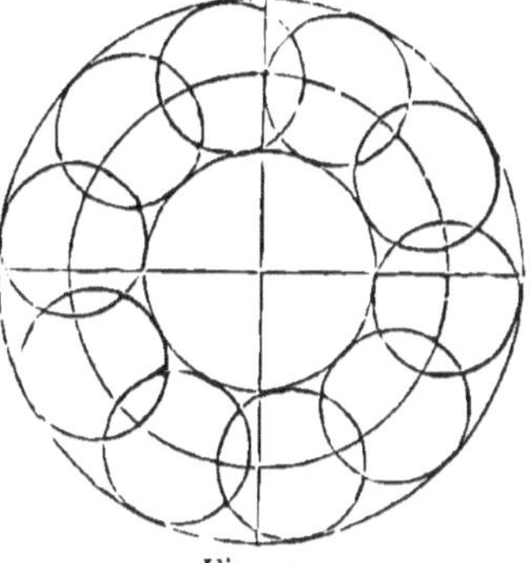

Fig. 51.

Differentiating the general equation of the circle with respect to a and b only we have

$$-(x-a)-(y-b)\frac{db}{da}=0$$

whence

$$\frac{db}{da}=-\frac{x-a}{y-b}$$

We may now substitute for b its value obtained from equation (1); or we may consider it as a function of a in that equation and substitute the value of the differential coefficient derived from it. This will give us

$$\frac{db}{da}=-\frac{a}{b}=-\frac{x-a}{y-b}$$

whence

$$(x-a)^2=\frac{(c^2-b^2)(y-b)^2}{b^2}$$

Substituting this value of $(x-a)^2$ in the general equation of the circle, we have

$$\frac{(c^2-b^2)(y-b)^2}{b^2}+(y-b)^2=R^2$$

from which we obtain

$$b=\frac{cy}{c\pm R}$$

and similarly

$$a=\frac{cx}{c\pm R}$$

Substituting these values of a and b in the general equation of the circle we have

$$\left(x-\frac{cx}{c\pm R}\right)^2+\left(y-\frac{cy}{c\pm R}\right)^2=R^2$$

whence

$$x^2+y^2=(c\pm R)^2$$

the equation of the envelope showing it to be twofold. The positive value of R gives a circle with a radius equal to $c+R$ circumscribing the system, and the negative value for R gives one that is inscribed within it.

(**123**) Let there be an ellipse in which the axes vary in length under the condition that the area of the ellipse shall be constant. This condition will be expressed by the equation

$$AB=c^2 \tag{1}$$

To find the envelope of this curve we put its equation under the form

$$\frac{x^2}{A^2}+\frac{y^2}{B^2}=1 \tag{2}$$

and differentiating with respect to A and B only we have

$$\frac{x^2}{A^3}+\frac{y^2}{B^3}\cdot\frac{dB}{dA}=0$$

or

$$\frac{1}{A}\cdot\frac{x^2}{A^2}=-\frac{1}{B}\cdot\frac{dB}{dA}\cdot\frac{y^2}{B^2}$$

But from equation (1) we obtain

$$\frac{1}{A}=-\frac{1}{B}\cdot\frac{dB}{dA}$$

whence

$$\frac{x^2}{A^2}=\frac{y^2}{B^2}=\frac{1}{2}$$

whence

$$A=x\sqrt{2} \text{ and } B=y\sqrt{2}$$

Substituting these values in equation (1) we have

$$2xy=AB=c^2$$

and

$$xy=\frac{c^2}{2}$$

which is the equation of a hyperbola referred to its center and asymptotes. The curve EF (Fig. 49) is then a hyperbola, and the axes of the ellipse are its asymptotes.

(124) Let AB (Fig. 48) and AC be the coordinate axes, and let the line de of a given length move in such a manner that its extremities shall be at all times in the axis. What is the equation of the envelope described by that line?

Call the length of the line c, and the distances Ad' and Ac' respectively b and a. Let $Am=x$ and $mn=y$, then the general equation of the line will be

$$\frac{x}{a}+\frac{y}{b}=1 \qquad (1)$$

we have also

$$a^2+b^2=c^2 \qquad (2)$$

Differentiating these equations with respect to a and b as variables we have

$$-\frac{db}{da}=\frac{a}{b}=\frac{b^2x}{a^2y}$$

or

$$\frac{y}{b}=\frac{b^2x}{a^3}$$

Substituting this value in equation (1) we have

$$\frac{x}{a}+\frac{b^2x}{a^3}=1$$

whence we obtain

$$a=\sqrt[3]{c^2x}$$

and similarly

$$b=\sqrt[3]{c^2y}$$

whence

$$a^2+b^2=(c^2x)^{\frac{2}{3}}+(c^2y)^{\frac{2}{3}}=c^2=(c^3)^{\frac{2}{3}}$$

from which

$$x^{\frac{2}{3}}+y^{\frac{2}{3}}=c^{\frac{2}{3}}$$

which is the equation of the envelope.

The first differential coefficient of this equation is

$$\frac{dy}{dx}=-\frac{x^{-\frac{1}{3}}}{y^{-\frac{1}{3}}}=-\frac{y^{\frac{1}{3}}}{x^{\frac{1}{3}}}$$

from which we learn that the curve is tangent to both coordinates.

(125) Suppose the line DC (Fig. 52) to revolve about the point D in the axis of abscissas, varying in length so that the extremity C shall be at all times in the axis of ordinates, required the envelope described by the line CE perpendicular to DC at the point C in the axis of ordinates.

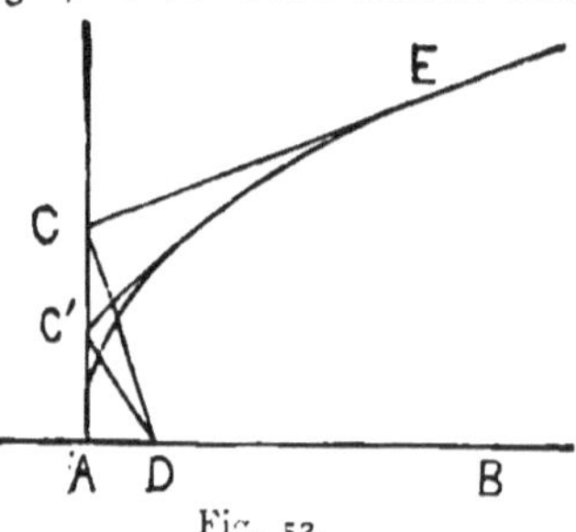

Fig. 52.

Representing the distance AD by c, and the tangent of the angle CDB by $-\frac{1}{a}$, its equation will be

$$y=-\frac{1}{a}(x-c)$$

in which if we make $x=o$ we have

$$y=\frac{c}{a}$$

for the distance from the origin at which the line DC intersects the axis of ordinates. And since the perpendicular passes through the same point, its equation will be

$$y=ax+\frac{c}{a} \qquad (1)$$

If we consider a in this equation as an independent variable, it will represent all the perpendiculars that can be drawn under the given condition.

Differentiating it with respect to a we have

$$x-\frac{c}{a^2}=0$$

whence

$$a=\sqrt{\frac{c}{x}}$$

and substituting this value of a in equation (1) we have

$$y\sqrt{\frac{c}{x}}=2c$$

whence

$$y^2=4cx$$

which shows the envelope to be a parabola of which D is the focus. It also demonstrates a well known property of the parabola, namely, if lines be drawn from the focus perpendicular to the tangent they will intersect it on the perpendicular to the axis through the vertex.

(126) Let AB and EO (Fig. 53) be the coordinate axes, and let CD be a line revolving between the lines AH and BK in such a manner that its extremities C and D shall always be in those lines, and the product of the distances CA and DB from the axis shall be a constant quantity. Required the equation of the envelope generated.

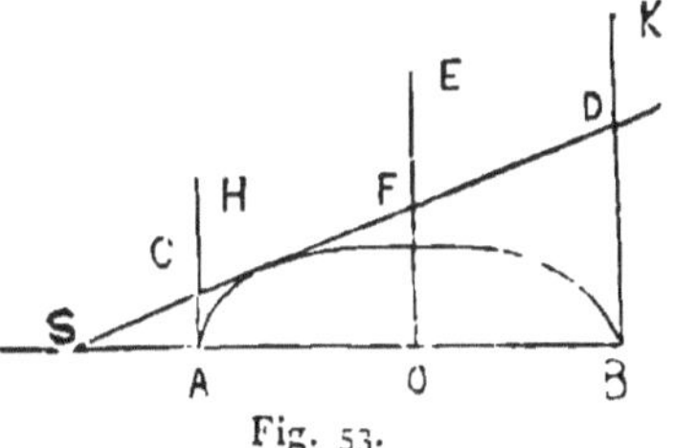

Fig. 53.

Let OA=OB=m, and AC . BD=c^2 Then producing the line DC until it meets the axis of abscissas at S, and making the tangent of BSD=a, we have

$$SB : BD :: SO : OF$$

or

$$\frac{OF}{a}+m : BD :: \frac{OF}{a} : OF$$

whence

$$BD=OF+am$$

and similarly

$$AC=OF-am$$

whence

$$BD \,.\, AC = c^2 = \overline{OF}^2 - a^2 m^2$$

or

$$\overline{OF}^2 = a^2 m^2 + c^2$$

But the equation of the line CD is

$$y = ax + b$$

in which b is the distance from the origin to the point where the line cuts the axis of ordinates, that is, the distance OF. Hence

$$y = ax + (a^2 m^2 + c^2)^{\frac{1}{2}} \qquad (:)$$

is the equation which, when a is variable, represents the line CD in every position it can assume under the given conditions.

Differentiating with respect to a, we have

$$x da + \frac{m^2 a da}{(a^2 m^2 + c^2)^{\frac{1}{2}}} = 0$$

whence

$$a = -\frac{x(a^2 m^2 + c^2)^{\frac{1}{2}}}{m^2}$$

$$= -\frac{c}{m} \cdot \frac{x}{(m^2 - x^2)^{\frac{1}{2}}}$$

which being substituted in equation (1) gives

$$y = -\frac{c}{m} \cdot \frac{x^2}{(m^2 - x^2)^{\frac{1}{2}}} + \left(\frac{c^2 x^2}{m^2 - x^2} + c^2\right)^{\frac{1}{2}}$$

whence

$$y(m^2 - x^2)^{\frac{1}{2}} = -\frac{c}{m} x^2 + cm$$

whence

$$my(m^2 - x^2)^{\frac{1}{2}} = c(m^2 - x^2)$$

whence

$$m^2 y^2 = m^2 c^2 - c^2 x^2$$

or

$$m^2y^2+c^2x^2=m^2c^2$$

which is the equation of an ellipse referred to its center as the origin, and whose semi-axes are m and c.

(127) The equation of the normal line to the parabola is

$$y-y'=-\frac{y'}{p}(x-x') \qquad (1)$$

in which x' and y' are the coordinates of the point in the curve from which the normal is drawn, and x and y are the variable coordinates of the normal itself.

If we consider x' and y' as variables, equation (1) will represent the entire system of normals which can be drawn to the parabola. To find the envelope of this system we find the relation between x' and y' from the equation of the parabola

$$y'^2=2px' \qquad (2)$$

and substitute in equation (1) the value of x', which gives

$$y-y'=-\frac{y'x}{p}+\frac{y'^3}{2p^2}$$

whence

$$2p^2(y-y')=-2pxy'+y'^3 \qquad (3)$$

Differentiating this equation with respect to y' only we have

$$-2p^2=-2px+3y'^2$$

whence

$$y'=\sqrt{\tfrac{2}{3}(px-p^2)}$$

Substituting this value in equation (3) we have

$$2p^2y-2p^2\sqrt{\tfrac{2}{3}(px-p^2)}=-2px\sqrt{\tfrac{2}{3}(px-p^2)}+[\tfrac{2}{3}(px-p^2)]^{\frac{3}{2}}$$

whence

$$2p^2y+2(px-p^2)(\tfrac{2}{3}(px-p^2))^{\frac{1}{2}}=[\tfrac{2}{3}(px-p^2)]^{\frac{3}{2}}$$

whence

$$p^2y=-[\tfrac{2}{3}(px-p^2)]^{\frac{3}{2}}=-(\tfrac{2}{3}p)^{\frac{3}{2}}(x-p)^{\frac{3}{2}}$$

whence

$$p^4y^2=\tfrac{8}{27}p^3(x-p)^3$$

or

$$y^2=\frac{8}{27p}(x-p)^3$$

which as we have seen (Art. 115) is the equation of the evolute of the parabola.

Hence all normal lines to the parabola are tangent to the evolute.

SECTION XV.

APPLICATION OF THE DIFFERENTIAL CALCULUS TO THE DISCUSSION OF CURVES.

THE CYCLOID.

(**128**) The cycloid is a curve described by any point in the circumference of a circle as it rolls along a straight line

If for example, the circle EFD (Fig. 54) should roll along the straight line AB, the point F, starting from the point A, would describe the cycloid AD′B, and the distance from A to B where the generating point again meets the line AB will be exactly equal to the circumference of the generating circle.

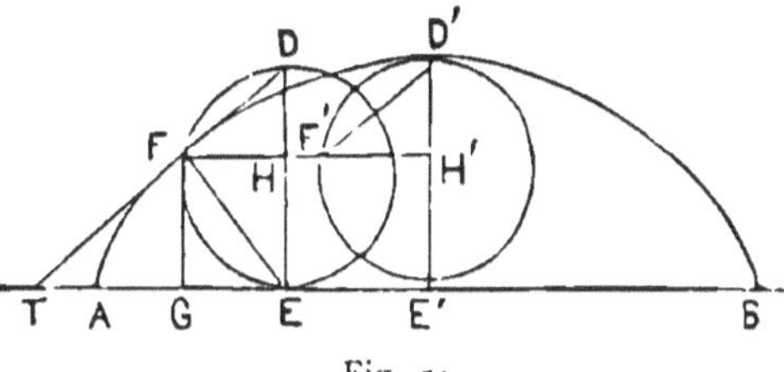

Fig. 54.

If we place the origin at A we shall have

$$AG=x \text{ and } FG=y$$

The arc FE will be equal to the line AE, and HE will be the versed sine of the same arc. Making $DE=2r$ we shall have

$$\overline{FH}^2=DH \,.\, HE=y(2r-y)$$

hence

$$FH=GE=\text{arc } FE-x=\sqrt{2ry-y^2}$$

whence

$$x=\text{ver. sin.}^{-1}y-\sqrt{2ry-y^2} \qquad (1)$$

which is the equation of the cycloid.

The line AB is called the base of the cycloid, and the line D′E′ perpendicular to the base at its middle point is the axis, and is equal to $2r$.

Since every negative value for y gives an imaginary value for x, the curve has no point below the base. If we make $y=2r$ we have

$$x=\text{ver sin.}^{-1}2r=\pi r$$

and every value for y greater than $2r$ gives an imaginary value for x; hence the greatest value of y is the diameter of the generating circle; and for all values of y between $2r$ and zero there will be a real value for x.

(**129**) We will now proceed, with the aid of the differential calculus, to investigate the properties of this curve in reference to its tangent, subtangent, normal, subnormal, curvature, evolute, etc.

Differentiating equation (1) we have

$$dx=\frac{rdy}{\sqrt{2ry-y^2}}-\frac{rdy-ydy}{\sqrt{2ry-y^2}}=\frac{ydy}{\sqrt{2ry-y^2}} \qquad (2)$$

Substituting this value of dx in the general formula for the subtangent (Art. 52) we have

$$TG=\frac{y^2}{\sqrt{2ry-y^2}}$$

and for the tangent (Art. 53)

$$TF=\sqrt{y^2+\frac{y^4}{2ry-y^2}}$$

For the subnormal (Art. 54)

$$GE=\sqrt{2ry-y^2}$$

and for the normal (Art. 55)

$$FE=y\sqrt{1+\frac{2ry-y^2}{y^2}}=\sqrt{2ry}$$

Since GE the subnormal is equal to $\sqrt{2ry-y^2}$, which is equal to $\sqrt{DH.HE}$, the point E of the subnormal for the point F of the curve, must be at the intersection of the vertical diameter of the corresponding generating circle with the base; and the normal line $=\sqrt{2ry}=\sqrt{DE.EH}$ *must be a chord of that circle joining these two points.*

The tangent being perpendicular to the normal will of course be the supplementary chord of the same circle. Hence to obtain the normal and tangent lines for any given point of the cycloid, construct the generating circle for the diameter D′E′ erected at the middle of the base, and through the given point draw the line FH′ parallel to the base intersecting the circle at F′. Join this point with the extremities of the diameter D′E′, and the line F′E′ will be parallel to the normal, and F′D′ will be parallel to the tangent. Hence lines parallel to these, through the given point will be the lines required.

If it is required to draw a tangent parallel to a given line, first draw a chord from D′ parallel to the given line, and through the point where it meets the circumference of the circle draw a line parallel to the base. The intersection of this line with the curve of the cycloid will be the point of tangency.

(**130**) From equation (2) we have

$$\frac{dy}{dx}=\frac{\sqrt{2ry-y^2}}{y}=\sqrt{\frac{2r}{y}-1} \qquad (3)$$

which becomes zero when $y=2r$, hence the tangent at the extremity of the axis is parallel to the base. If we make $y=0$ we have

$$\frac{dy}{dx}=\infty$$

hence the tangent at the base is perpendicular to it.
Differentiating equation (3) we have

$$\frac{d^2y}{dx}=-\frac{\frac{2rdy}{y^2}}{2\sqrt{\frac{2r}{y}-1}}=-\frac{\frac{rdy}{y^2}}{\frac{dy}{dx}}=-\frac{rdx}{y^2}$$

hence

$$\frac{d^2y}{dx^2}=-\frac{r}{y^2}$$

This second differential coefficient being essentially negative, shows that the curve is everywhere concave toward the base.

(**131**) The formula for the radius of curvature (Art. 105) gives in this case,

$$R=\frac{\left(dx^2+\frac{2rdx^2}{y}-dx^2\right)^{\frac{3}{2}}}{\frac{rdx^3}{y^2}}=\frac{\left(\frac{2rdx^2}{y}\right)^{\frac{3}{2}}}{\frac{rdx^3}{y^2}}=2^{\frac{3}{2}}r^{\frac{1}{2}}y^{\frac{1}{2}}$$

or

$$R=2\sqrt{2ry}$$

But we have found (Art. 129) the normal to be equal to $\sqrt{2ry}$; hence the radius of curvature at any point is equal to twice the normal at that point. Thus at A the radius of curvature is nothing, while at D′ it is equal to $2D'E'=4r$.

(**132**) The equation of the evolute will be found by the rule given in Art. 114.

In the equations of condition (Art. 105)

$$x-a=-\frac{dy}{dx}(y-b) \tag{2}$$

and

$$y-b=-\frac{dx^2+dy^2}{d^2y} \tag{3}$$

Substitute the values of $-\frac{dy}{dx}$ and $\frac{d^2y}{dx^2}$ just found from the equation of its cycloid, and then, by means of that equation eliminate x and y.

Thus

$$x-a=-\frac{\sqrt{2ry-y^2}}{y}(y-b)$$

and

$$y-b=\frac{dx^2+dx^2\left(\frac{2r}{y}-1\right)}{\frac{dx^2r}{y^2}}$$

whence

$$y-b=2y \text{ and } x-a=-2\sqrt{2ry-y^2}$$

or

$$y=-b \text{ and } x=a-2\sqrt{-2rb-b^2}$$

Substituting these values of x and y in equation (1) (Art. 128), we have

$$a-2\sqrt{-2rb-b^2}=\text{ver. sin.}^{-1}-b-\sqrt{-2rb-b^2}$$

or

$$a=\text{ver. sin.}^{-1}-b+\sqrt{-2rb-b^2} \qquad (4)$$

which is the equation of the evolute.

(**133**) For all values of b that are positive a is imaginary, hence no part of the curve is above the base of the involute. For all negative values of b greater than $2r$, a is also imaginary, hence if we draw A′B′ (Fig. 55) parallel to the base at a distance below it equal to $2r$, the evolute will lie between that line and the base. If we make $b=-2r$, a becomes equal to the arc whose versed sine is $-b$,

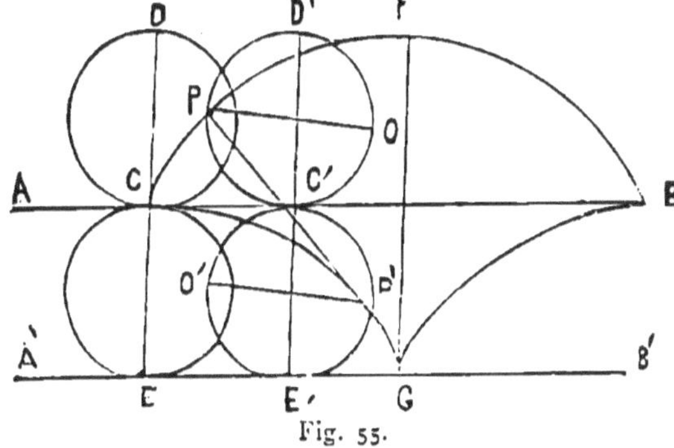

Fig. 55.

that is half the circumference of the generating circle. Hence the point G where the evolute meets the line A′B′ is in the prolonged axis of the involute. If we make $b=0$, a also becomes equal to zero, and hence the evolute passes through the origin at A, and also the extremity of the base at B. For ver. $\sin.^{-1}o$ may be zero, or it may be a whole circumference.

If we differentiate equation (4) we have

$$da=-\frac{rdb}{\sqrt{-2rb-b^2}}-\frac{rdb+bdb}{\sqrt{-2rb-b^2}}=-\frac{(2r+b)db}{\sqrt{-2rb-b^2}}$$

or

$$\frac{db}{da}=-\frac{\sqrt{-2rb-b^2}}{2r+b}$$

showing that at the points C and B where $b=o$ the base of the involute is tangent to the evolute. Also since

$$-\frac{\sqrt{-2rb-b^2}}{2r+b}=\frac{b}{\sqrt{-2rb-b^2}}$$

if we make $b=-2r$ we have

$$\frac{db}{da}=\infty$$

showing the tangent to the evolute at G is perpendicular to the line A′B′.

Squaring the value of $\frac{db}{da}$, and differentiating, we have

$$\frac{db^2}{da^2}=-\frac{b}{2r+b} \text{ and } \frac{d^2b}{da^2}=-\frac{r}{(2r+b)^2}$$

which is essentially negative, and since every real value of b is also negative, the curve is everywhere convex to the base of the cycloid.

(**134**) These circumstances, together with the form of the equation of the evolute, lead us to suppose it to be an equal cycloid, but for certainty we will transfer the origin to G, and the coordinate axes to EG and GF respectively parallel to the first. Calling the new coordinates x' and y' we have

$$a=x'+m \text{ and } b=y'+n$$

m and n being the coordinates of the new origin referred to the original axes. Then

$$m=\text{ver. sin.}^{-1}2r \text{ and } n=-2r$$

whence

$$a=x'+\text{ver. sin.}^{-1}2r \text{ and } b=-(2r-y')$$

Substituting these values of a and b in equation (4) we have

$$x'+\text{ver. sin.}^{-1}2r=\text{ver. sin.}^{-1}(2r-y')+\sqrt{2r(2r-y')-(2r-y')^2}$$

but

$$\text{ver. sin.}^{-1}2r-\text{ver. sin.}^{-1}(2r-y')=\text{ver. sin.}^{-1}y'$$

hence

$$x'=-\text{ver. sin.}^{-1}y'+\sqrt{2ry'-y'^2}$$

which is the equation of the curve CG, the values of x' being the same as those of x in equation (1) (Art. 128), except that they are negative as they should be, since the values of x' are reckoned in a contrary direction from those of x; and the curve CG is equal to the curve CF, but reversed in position with reference to the origin.

Since the curve CG is equal to FG (Art. 113) the length of the cycloid is equal to four times the diameter of the generating circle.

(**135**) The character of the evolute of the cycloid may be demonstrated geometrically thus:

Let us suppose two right lines AB and A′B′ (Fig. 55) to be drawn parallel to each other, and two circles to be described on the diameters DC and CE, each equal to the distance between the two parallel lines and tangent to each other at the point C. If now we suppose each circle to roll along the line on which it stands, at the same rate, so that they are at all times tangent to each other, then the point C of the upper circle will describe the first half of a cycloid CPF, while the same point C of the lower circle will describe the last half of an equal cycloid CP′G.

Suppose the two circles to have arrived at the point C′ in the line AB, and that P is a point in the upper curve. The diameter DC of the upper circle will have assumed the position PO, and the diameter CE of the lower circle will have assumed the position O′P′ parallel to it; and P′ will be the generating point of the lower cycloid.

Draw the chord PC′ and it will be normal to the upper cycloid (Art. 129). Draw also the chord C′P′, and it will be tangent to the lower cycloid at the point P′ (Art. 129). Now since PO and O′P′ are parallel, these two chords and the corresponding arcs are equal, and hence the angles PC′D′ and P′C′E′ are equal; and since D′E′ is a straight line P′C′P is a straight line also, normal to the upper curve and tangent to the lower one. Hence the lower cycloid is the evolute of the upper one.

(136) The equation of the evolute may also be obtained by considering it as the envelope of the normals drawn to the curve.

The general equation of the normal to the cycloid is

$$y-y' = -\frac{y'}{\sqrt{2ry'-y'^2}}(x-x') \qquad (1)$$

in which x' and y' are the coordinates of that point of the cycloid to which the normal is drawn; and x and y the general coordinates of the normal line. If we make x' and y' variables, still retaining their relative values, as in the equation of the cycloid, the equation (1) will represent the whole system of normals that can be drawn to the curve. If now we eliminate one and make the differentials of the equation with respect to the other equal to zero, then (Art. 120) by eliminating that we shall have an equation which will be that of the envelope of the normals, and also the evolute of the cycloid.

Substituting for x' in equation (1) its value taken from the equation of the curve (Art. 128) we have

$$y-y'=-\frac{y'}{\sqrt{2ry'-y'^2}}(x-\text{ver. sin.}^{-1}y'+\sqrt{2ry'-y'^2})$$

or

$$y-y'=\frac{-y'x+y'\text{ver. sin.}^{-1}y'}{\sqrt{2ry'-y'^2}}-y'$$

whence

$$y\sqrt{2ry'-y'^2}-y'\text{ver. sin.}^{-1}y'+xy'=0 \qquad (2)$$

Differentiating this equation with respect to y' we have

$$y\frac{r-y'}{\sqrt{2ry'-y'^2}}-\text{ver. sin.}^{-1}y'-y'\frac{r}{\sqrt{2ry'-y'^2}}+x=0$$

Substituting in this equation for ver. sin.$^{-1}y'$ its value taken from the equation of the cycloid, and multiplying by $\sqrt{2ry'-y'^2}$, we have

$$y(r-y')-(x'+\sqrt{2ry'-y'^2})\sqrt{2ry'-y'^2}-y'r+x\sqrt{2ry'-y'^2}=0$$

or

$$y(r-y')=x'\sqrt{2ry'-y'^2}+2ry'-y'^2+y'r-x\sqrt{2ry'-y'^2}$$

or

$$y(r-y')=-(x-x')\sqrt{2ry'-y'^2}+3ry'-y'^2$$

but

$$x-x'=-\frac{y-y'}{y'}\sqrt{2ry'-y'^2}$$

hence

$$y(r-y')=\frac{y-y'}{y'}(2ry'-y'^2)+3ry'-y'^2$$

clearing of fractions and multiplying we have

$$ryy'-yy'^2=2ryy'-2ry'^2-yy'^2+y'^3+3ry'^2-y'^3$$

whence

$$ryy'=-ry'^2 \text{ or } y=-y'$$

Substituting this value of y' in equation (2) we have

$$y\sqrt{-2ry-y^2}+y\text{ ver. sin.}^{-1}-y-xy=0$$

or

$$x=\text{ver. sin.}^{-1}-y+\sqrt{-2ry-y^2}$$

which is the equation of the envelope of the normals, and also of the evolute of the cycloid, as in Art. 132; for substituting the variables a and b for x and y, the equations are identical.

THE LOGARITHMIC CURVE.

(137) The logarithmic curve is one in which one of the coordinates is the logarithm of the other.

Its equation is

$$x = \text{Log.}\, y$$

If we represent the base of the system by a the equation may be written

$$y = a^x$$

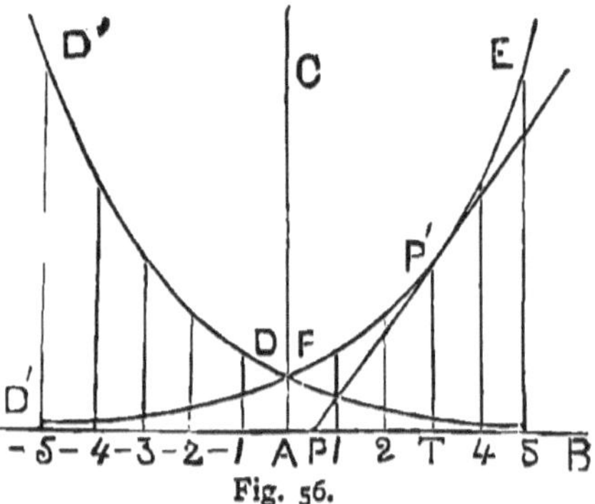

Fig. 56.

The curve may be constructed by laying off on AB (Fig. 56) the axis of logarithms, the numbers 1, 2, 3, 4, etc., on both sides of the origin, and laying off on the corresponding ordinates, or on AC the axis of numbers, the corresponding powers of a.

When $x=0$ then $y=1$, whatever may be the value of a, and hence all logarithmic curves will intersect the axis of numbers at a distance from the origin equal to 1.

If a is greater than 1, and x positive, y will increase as x increases, and there will be a real value of y for every value of x as in the curve DE.

If x is negative, then the value of y is fractional, and decreases as x increases negatively, but y will not become zero until $x = -\infty$.

If y is negative, there is no corresponding value of x, and hence the curve can never pass below the axis AB.

If a is less than 1, then y will diminish as x increases positively, and becomes zero when $x=\infty$; but y increases for negative values of x, and the curve has a position the reverse of the first as DD″ in the figure.

(**138**) If we differentiate the equation

$$y=a^x$$

we have

$$\frac{dy}{dx}=a^x\frac{1}{m}=y\frac{1}{m}$$

and

$$\frac{d^2y}{dx^2}=a^x\cdot\frac{1}{m^2}=y\frac{1}{m^2}$$

If we make $y=0$ we have $\frac{dy}{dx}=0$, hence the tangent for that value of y is the axis of abscissas; and since $y=0$ gives $x=-\infty$ the axis of abscissas is an asymptote to the curve (Art. 88). But since $y=\infty$ gives $x=\infty$, and also $\frac{dy}{dx}=\infty$, the curve has no tangent parallel to the axis of ordinates except at an infinite distance. The *sign* of the second differential coefficient shows that the curve is at all times convex toward the axis of abscissas.

The subtangent $\text{PT}=\frac{dx}{dy}y=\text{M}$; hence the subtangent is constant and equal to the modulus of the system of logarithms, to which the curve belongs.

In the Naperian system the modulus is 1, and in this case PT and DA are equal.

(**139**) We will now investigate the curve whose equation is

$$y=x\log. x$$

Every value for x gives a single value for y.

If x is less than 1 the value of y is negative. (1)

If x is greater than 1, y is positive. (2)

If $x=0$, or $x=1$, $y=0$. (3)

If x is negative, y is imaginary. (4)

If we differentiate the equation we have

$$\frac{dy}{dx} = \log. x + 1 \tag{5}$$

$$\frac{d^2 y}{dx^2} = \frac{1}{x} \tag{6}$$

Making $\frac{dy}{dx} = 0$ we have

$$\log. x = -1 \text{ or } x = e^{-1} = \frac{1}{e} = \frac{1}{2.7182} \tag{7}$$

which corresponds to a minimum as shown by the positive value of $\frac{d^2 y}{dx^2}$. (8)

When $x = 0$, $\frac{dy}{dx} = \infty$. (9)

When $x = 1$, $\frac{dy}{dx} = 1$. (10)

Since y is negative between $x = 0$ and $x = 1$ and then positive, while $\frac{d^2 y}{dx^2}$ is always positive, the curve is concave toward the axis of abscissas between $x = 0$ and $x = 1$, and afterwards convex. (11)

Hence the curve begins at the origin (Fig. 57) and intersects the axis of abscissas at D, making AD $= 1$ (3). The tangent to the curve at D makes an angle of 45° with the axis of abscissas (10), while at A the axis of ordinates is tangent (9). At the point E, whose abscissa is $\frac{1}{2.7182}$, the tangent to the curve is parallel to the axis of abscissas (7), and the ordinate is at a minimum (or negative maximum) (8). Between A and D the curve is below the axis of abscissas (1), and concave to it (11); and

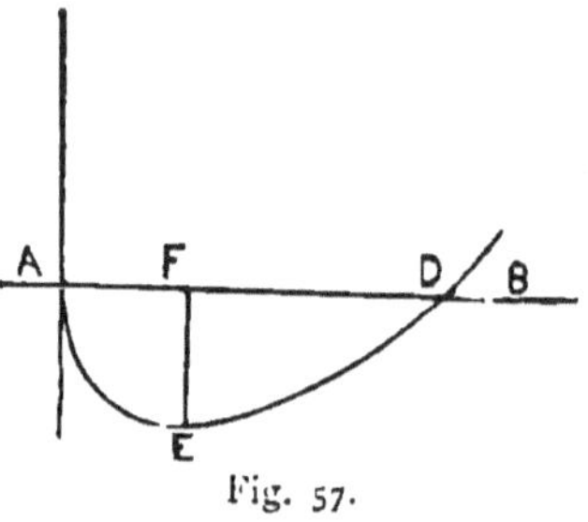

Fig. 57.

beyond D the curve lies entirely above the axis of abscissas and is convex to it. Since $x=\frac{1}{2.7182}$ gives $y=-x$, we have

$$FE=AF$$

(140) We will next take the equation

$$y=e^{-\frac{1}{x}}=\frac{1}{e^{\frac{1}{x}}}$$

Every value of x gives a real positive value for y, and hence there can be no negative value of y. (1)

If $x=0$, $y=0$, and hence the curve passes through the origin. (2)

If x is negative we have $y=e^{\frac{1}{x}}$, in which if $x=0$, y will become infinite. (3)

So that $x=0$ gives two values for y, according as x approaches zero from the positive or negative side. (4)

If x be negative and increase in value, that of y will approach more nearly to 1, which it will reach when $x=-\infty$. (5)

If x be positive, and increasing the value of y approaches more nearly to 1, which it reaches when $x=\infty$. (6)

Hence the curve will be as in Fig. 58, in which AB and AC are the axes of coordinates, and DE a line parallel to AB at a distance from it equal to 1. It will pass through the origin A (2), extend indefinitely in a positive and negative direction, and the line DE will be an asymptote to both branches (5), (6). The axis of ordinates will also be an asymptote in the positive direction (3) (Art. 88). As the branch DC of the curve extends to an infinite distance in both directions, it has no connection with

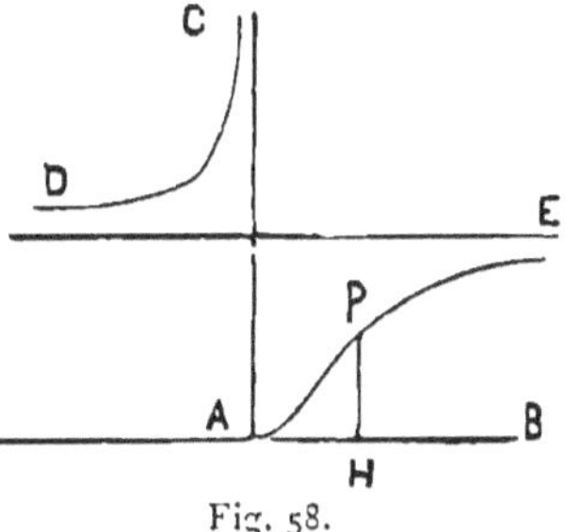

Fig. 58.

the branch AE, which commences at the origin and is infinite at the other extremity. There are, in fact, two curves, one answering to the positive, and the other to the negative value of x.

If we differentiate the equation $y=e^{-\frac{1}{x}}$ we have

$$\frac{dy}{dx}=\frac{e^{-\frac{1}{x}}}{x^2}$$

and

$$\frac{d^2y}{dx^2}=\frac{e^{-\frac{1}{x}}(1-2x)}{x^4}$$

Since

$$\frac{e^{-\frac{1}{x}}}{x^2}=\frac{1}{x^2e^{\frac{1}{x}}}$$

we shall have

$$\frac{dy}{dx}=0$$

when either $x=0$ or $x=\infty$, hence the axis of abscissas is tangent at the origin, and parallel to the tangent at an infinite distance in either direction; in which case $y=1$ (5) (6).

For all negative values of x, $\frac{d^2y}{dx^2}$ is positive, and hence the branch DC is convex to the axis of abscissas. For all positive values of x less than $\frac{1}{2}$, $\frac{d^2y}{dx^2}$ is also a positive, showing that between A and H the curve is convex to the axis of abscissas, while at the point H, where $x=\frac{1}{2}$, the value of $\frac{d^2y}{dx^2}$ changes from positive to negative passing through zero, showing that at P the curve ceases to be convex, and becomes concave toward the axis of abscissas. This is called an inflexion.

SECTION XVI.

SINGULAR POINTS.

(141) Singular points of a curve are those at which there exists some remarkable property not common to other points of it. Such, for example, as the maximum or minimum value of the ordinates or abscissas, points of inflexion, conjugate points, cusps, etc.

In many cases these points are easily discovered by the aid of the differential calculus, as will be seen by the following examples.

MAXIMA AND MINIMA.

(142) If we differentiate the equation

$$y=3+2(x-4)^4$$

we shall have

$$\frac{dy}{dx}=8(x-4)^3$$

$$\frac{d^2y}{dx^2}=24(x-4)^2$$

$$\frac{d^3y}{dx^3}=48(x-4)$$

$$\frac{d^4y}{dx^4}=48$$

Here we find that $x=4$ will reduce the first differential

coefficient to zero, showing that the tangent to the curve is parallel to the axis of abscissas (Art. 36), and hence the value of the ordinate *may* be a maximum or minimum. But since the second differential coefficient is always positive except when it is zero, the first must be an increasing function, and hence at zero must be passing from negative to positive, and the value of y must be changing from a diminishing to an increasing one. So that there is a minimum when $x=4$, as shown in Fig. 59. We infer the same thing from the sign of the fourth differential coefficient (Art. 29).

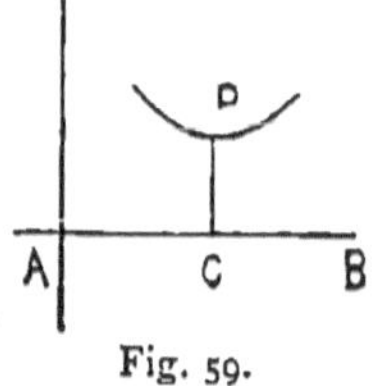

Fig. 59.

If we take the equation

$$y=2-2(x-2)^4$$

we shall have

$$\frac{dy}{dx}=-8(x-2)^3$$

$$\frac{d^2y}{dx^2}=-24(x-2)^2$$

$$\frac{d^3y}{dx^3}=-48(x-2)$$

$$\frac{d^4y}{dx^4}=-48$$

Since $x=2$ reduces the first differential coefficient to zero the tangent at that point is parallel to the axis of abscissas; and since the fourth differential coefficient (the first that has real value for $x=2$) is negative, the value of y at that point must be a maximum as in the figure. Since $\frac{d^2y}{dx^2}$ is at all times negative, except when $x=2$, the curve will be concave toward the axis of abscissas for all positive values of y (Art. 91).

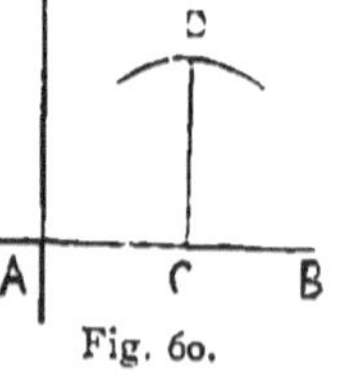

Fig. 60.

POINTS OF INFLEXION.

(143) A point of inflexion is one in which the radius of curvature changes from one side to the other of the curve so that it will be convex on one side of the point of inflexion, and concave on the other towards any line not passing through the point itself, and this will, of course, be true for the axis of abscissas, and hence at such a point the second differential coefficient will change its sign. If the point of inflexion should be in the axis of abscissas, both parts of the curve would be convex or concave to it, but the second differential coefficient will still change its sign (Art. 93). Now in order that the sign should be changed, the function must pass through zero or infinity, and hence the equations

$$\frac{d^2y}{dx^2}=0 \text{ and } \frac{d^2y}{dx^2}=\infty$$

will give all the points of inflexion in any curve in which there may be such points.

(144) Let us now take the equation

$$y=1+3(x-2)^3$$

whence

$$\frac{dy}{dx}=9(x-2)^2$$

and

$$\frac{d^2y}{dx^2}=18(x-2)$$

$$\frac{d^3y}{dx^3}=18$$

In this case every value of y gives one for x, and *vice versa*, hence the curve has no limit. When $x=2$, then $\frac{dy}{dx}=0$ and $y=1$, so that if we make AC$=2$ and CD$=1$ (Fig. 61), the tangent at D will be parallel to the axis AB. But $x=2$ reduces $\frac{d^2y}{dx^2}$ to zero, also indicating that

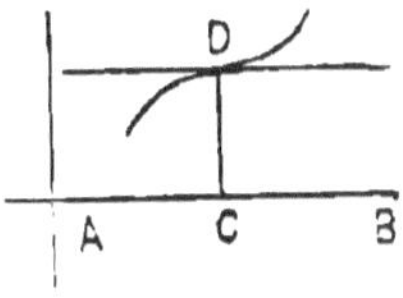

Fig. 61.

there *may* be a point of inflexion, hence we resort to the value of $\frac{d^3y}{dx^3}$ which is a positive constant. From this we learn that at zero $\frac{d^2y}{dx^2}$ is an increasing function, hence it must pass from negative to positive, showing that at the same point $\frac{dy}{dx}$ passes from a decreasing to an increasing function, and hence does not change its sign, but remains positive both before and after the zero point; and this shows that the value of y is an increasing function both before and after the same point. There is, therefore, no maximum nor minimum for it at that point.

Since $\frac{d^2y}{dx^2}$ changes its sign at $x=2$ from negative to positive, the curve will be concave toward the axis of abscissas when $x<2$, and convex when $x>2$, so that at the point D where $x=2$ the curvature changes its direction and there is an inflexion.

(**145**) If we take the same equation, and make the last term negative, we shall have

$$\frac{dy}{dx}=-9(x-2)^2$$

$$\frac{d^2y}{dx^2}=-18(x-2)$$

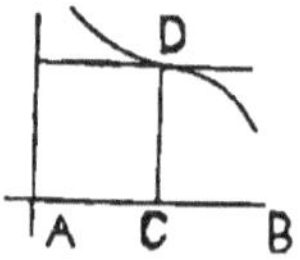

Fig. 62.

The point D where $x=2$ will still be the point of inflexion, but since $\frac{dy}{dx}$ is negative for all values of x except $x=2$, the curve will approach the axis of abscissas for all positive values of y, except $y=1$, and since $\frac{d^2y}{dx^2}$ is positive for $x<2$, and negative for $x>2$, the curve will be convex toward AB between A and C, and concave afterwards, as in Fig. 62.

The first differential coefficient being zero when $x=2$, it follows that the tangent at that point will be parallel to the axis of abscissas (Art. 36), and hence the curve will pass from one side of the tangent to the other at the point of tangency, and will be convex to the tangent on both sides of it.

(**146**) If we take the equation

$$y=2+2(x-2)^{\frac{3}{5}}$$

we have

$$\frac{dy}{dx}=\frac{6}{5(x-2)^{\frac{2}{5}}}$$

$$\frac{d^2y}{dx^2}=-\frac{12}{25(x-2)^{\frac{7}{5}}}$$

If we make $x=2$ we have

$$y=2,\quad \frac{dy}{dx}=\infty,\quad \frac{d^2y}{dx^2}=\infty$$

and hence at the point D where $x=2$ the tangent will be perpendicular to the axis of abscissas (Fig. 63), and since $\frac{dy}{dx}$ is positive for other values of x, the curve will leave the axis of abscissas, for all positive values of y as x increases. And since $\frac{d^2y}{dx^2}$ changes its sign from positive to negative in passing through infinity where $x=2$, the curve will be convex toward the axis of abscissas for $x<2$, and concave for $x>2$, and at $x=2$ there will be an inflexion.

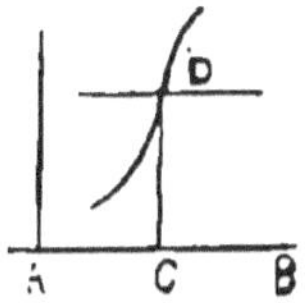

Fig. 63.

(**147**) If in the same equation we make the last term negative we have

$$y=2-2(x-2)^{\frac{3}{5}}$$

and

$$\frac{dy}{dx}=-\frac{6}{5(x-2)^{\frac{2}{5}}}$$

$$\frac{d^2y}{dx^2}=\frac{12}{25(x-2)^{\frac{7}{5}}}$$

and the conditions will be changed so that the curve will be reversed. It will now approach the axis of abscissas, and the second differential coefficient will change its sign from negative to positive in passing through infinity where $x=2$; the curve will be concave for $x<2$, and convex for $x>2$.

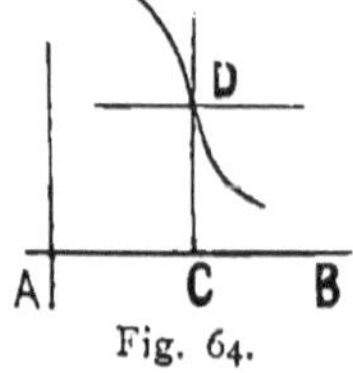

Fig. 64.

The point D (Fig. 64) will still be a point of inflexion, and the tangent will be perpendicular to AB.

(**148**) If we take the equation

$$y=(x-2)^3$$

we have

$$\frac{dy}{dx}=3(x-2)^2$$

and

$$\frac{d^2y}{dx^2}=6(x-2)$$

which all reduce to zero when $x=2$.

This shows that the curve meets the axis of abscissas at the point where $x=2$, and that this axis is tangent to it there. And since the second differential coefficient will have the same sign as y (both being the same as that of $x-2$), it will change from negative to positive at the point where $x=2$, showing an inflexion there, and that the curve is convex to the axis of abscissas on both sides of it.

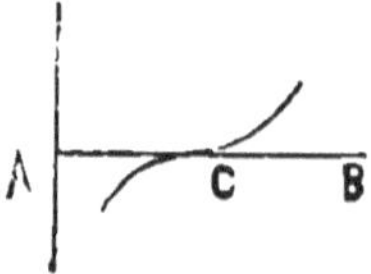

Fig. 65.

CUSPS.

(**148**) A cusp is a curve consisting of two branches starting from a common point in the same direction, and immediately diverging from a common tangent. They are of two kinds, namely: Those in which the branches are on different sides of the tangent, which are cusps of the first order: and those in which the branches are both on the same side of the tangent, which are cusps of the second order.

The following are examples of the first order.

Let

$$y = 1 + 3(x-1)^{\frac{2}{3}}$$

then

$$\frac{dy}{dx} = 2(x-1)^{-\frac{1}{3}}$$

and

$$\frac{d^2y}{dx^2} = -\frac{2}{3(x-1)^{\frac{4}{3}}}$$

If we make $x=1$ we have

$$y=1, \quad \frac{dy}{dx} = \infty, \text{ and } \frac{d^2y}{dx^2} = -\infty$$

For every value of $x<1$, $\frac{dy}{dx}$ is negative, and positive for every value greater. The curve, therefore, approaches the axis of abscissas, in the first case, and recedes from it in the second (y being positive), which indicates a minimum, while the tangent at that point is perpendicular to the axis of abscissas; and since $\frac{d^2y}{dx^2}$ is always negative the curve is concave toward the same axis.

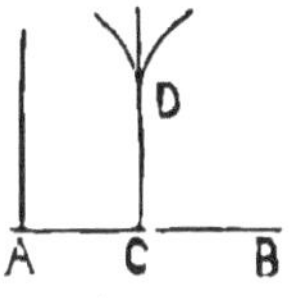

Fig. 66.

Every value of y less than 1 gives an imaginary value for x, while every value greater than 1 gives two values for x, one less and one as much greater than 1. Hence the curve has two equal branches commencing at D (Fig. 66), where they have a common tangent.

(150) If we make the last term negative, the signs of $\frac{dy}{dx}$ and $\frac{d^2y}{dx^2}$ will be reversed, and the first will change from positive to negative as x passes from $x<1$ to $x>1$; while at $x=1$ the tangent is still perpendicular to the axis of abscissas. Any value of y greater than 1 will give an imaginary value for x, while every value less than 1 will give two real values for x equally distant from the point C where $x=1$ (Fig. 67). The sign of $\frac{d^2y}{dx^2}$ being now always positive (except at $x=1$) shows that both branches of the curve are convex toward the axis of abscissas. These are then cusps of the first order.

Fig. 67.

(151) If we differentiate the equation

$$y=2\pm(x-1)^{\frac{3}{2}} \qquad (1)$$

we have

$$\frac{dy}{dx}=\pm\tfrac{3}{2}(x-1)^{\frac{1}{2}}$$

$$\frac{d^2y}{dx^2}=\pm\tfrac{3}{4}(x-1)^{-\frac{1}{2}}$$

We see from equation (1) that when $x=1$, $y=2$, and when $x<1$, y is imaginary, while when $x>1$, y has two values, one greater than 2 and the other as much less; so that if DC (Fig. 68) be drawn perpendicular to AB, making $DC=2$, the curve will commence at D and be symmetrical about the line DE, and since $\frac{dy}{dx}=0$ for the point D, the line DE will be tangent to both branches. Since for every other value of x, $\frac{dy}{dx}$ has one negative and one equal positive value, one branch of the curve will approach the axis of abscissas, and the other recede from it at an equal rate. And since for every value of $x>1$, $\frac{d^2y}{dx^2}$ has two equal values with contrary signs, the positive corresponding with the greatest value of y, we infer that the

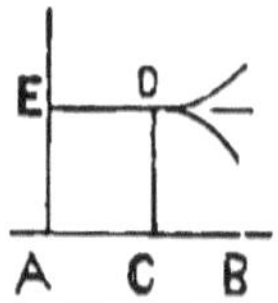

Fig. 68.

upper branch of the curve is convex, and the lower branch concave, to the axis of abscissas, and that the curve is a cusp of the first order.

(152) If we change the sign of the last term and make the equation

$$y=2\pm(1-x)^{\frac{3}{2}}$$

we have

$$\frac{dy}{dx}=\mp\tfrac{3}{2}(1-x)^{\frac{1}{2}}$$

$$\frac{d^2y}{dx^2}=\pm\frac{3}{4(1-x)^{\frac{1}{2}}}$$

and the curve will be similar, but reversed in position as in Fig. 69.

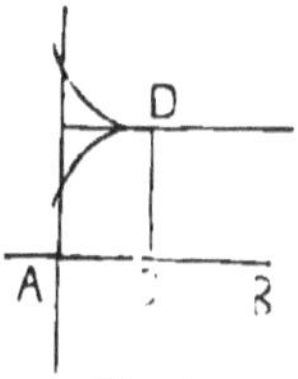

If $x>1$, y will be imaginary.

If $x=0$, $y=3$ and $y=1$.

If $y=0$, $x=1-\sqrt[3]{4}$. Since $\frac{d^2y}{dx^2}$ is both positive and negative when $x<1$ there is no maximum nor minimum value for y.

Fig. 69.

(153) The curve represented by the equation

$$(y-x^2)^2=x^5$$

contains a cusp of the second order, as well as some other singular properties.

Solving this equation we have

$$y=x^2\pm x^{\frac{5}{2}} \qquad (1)$$

and by differentiation we have

$$\frac{dy}{dx}=2x\pm\tfrac{5}{2}x^{\frac{3}{2}}$$

and

$$\frac{d^2y}{dx^2}=2\pm\frac{15}{4}x^{\frac{1}{2}}$$

From equation (1) we find that the curve passes through

the origin, and does not extend to the negative side of the axis of ordinates. Every positive value for $x<1$ gives two real positive values for y, while $x=1$ gives one positive value for y and one equal to zero. Hence the curve has two branches, both of which pass through the origin, and one intersects the axis of abscissas at a distance from the origin equal to 1.

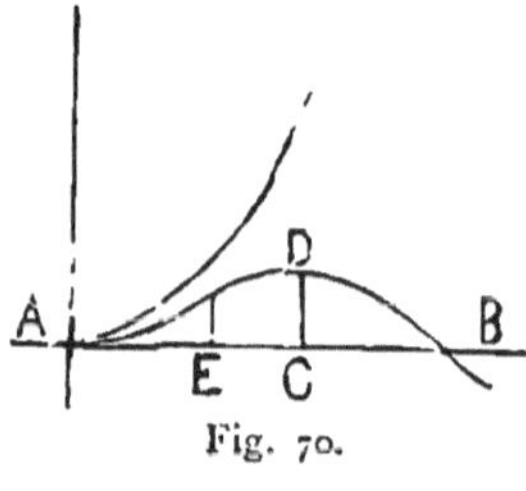

Fig. 70.

If we make $\frac{dy}{dx}=0$ we have $x=0$ and $x=\frac{16}{25}$. Hence there are two points in which the tangent to the curve is parallel to the axis of abscissas; at the origin where the axis itself is tangent and at the point D (Fig. 70) whose abscissa is $x=\frac{16}{25}$; and as the value of x at this point derived from the equation $\frac{dy}{dx}=0$ corresponds to the minus sign in equation (1), the point of tangency is on the lower branch of the curve.

The second differential coefficient has two values $2+\frac{15}{4}x^{\frac{1}{2}}$ and $2-\frac{15}{4}x^{\frac{1}{2}}$, of which the first belongs to the upper branch of the curve and is always positive, while the second is positive so long as $\frac{15}{4}x^{\frac{1}{2}}$ is less than 2; that is so long as x is less than $\frac{64}{225}$, which makes $\frac{d^2y}{dx^2}=0$. After that it becomes negative; showing that the lower branch of the curve is convex to the axis of abscissas, as far as the point whose abscissa is $AE=\frac{64}{225}$, and at this point there is an inflexion, the curve becoming concave to the axis of abscissas as long as y is positive and convex afterward. Hence at the origin there is a cusp of the second species.

CONJUGATE POINTS.

(154) Conjugate points are those single points which are isolated from the curve, but will satisfy the equation.

If we differentiate the equation

$$y=\pm\sqrt{\frac{x^2(x-b)}{a}} \qquad (1)$$

we have

$$\frac{dy}{dx}=\pm\frac{3x-2b}{2\sqrt{a(x-b)}}$$

$$\frac{d^2y}{dx^2}=\pm\frac{3x-4b}{4a^{\frac{1}{2}}(x-b)^{\frac{3}{2}}}$$

If we make $x=0$ in equation (1) we have $y=0$, but any other value of x less than b will make y imaginary. Hence while the origin will satisfy the equation, that point is isolated, having no connection with the curve. We also see that $x=0$ will give

$$\frac{dy}{dx}=\pm\frac{-b}{\sqrt{-ab}}$$

which is imaginary as it should be, since at that point the curve can have no tangent.

If we make $x=b$, we have

$$\frac{dy}{dx}=\infty$$

showing that the tangent at that point is perpendicular to the axis of abscissas, while the value of y is zero. As every positive value of $x>b$ gives two equal values for y with opposite signs, the curve is symmetrical about the axis of abscissas, and as the value of $\frac{dy}{dx}$ has the same sign as y, the curve departs from that axis in both directions. If we make x negative the value of y becomes imaginary; showing that

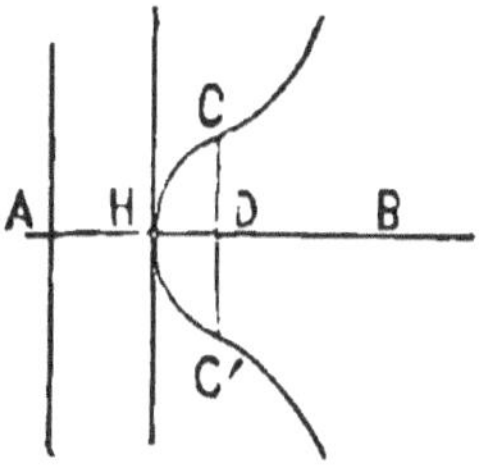

Fig. 71.

the curve does not extend to the negative side of the axis of ordinates.

If we make $\frac{d^2y}{dx^2}=0$, we have

$$x=\frac{4b}{3}$$

showing that at the points C and C′ (Fig. 71) which lie in the ordinate drawn through D at a distance from the origin equal to $\frac{4}{3}b$, the curve has an inflexion in each branch, since for that value of x we have

$$y=\pm\tfrac{4}{3}b\sqrt{\frac{b}{3a}}$$

If we make $x<\frac{4b}{3}$, the second differential coefficient will have a sign contrary to that of y. If $x>\frac{4b}{3}$ the signs will be the same. Hence between H and D the curve is concave toward the axis of abscissas, and convex beyond D, which also shows an inflexion.

If we make $\frac{dy}{dx}=0$ we have $3x=2b$, or

$$x=\frac{2b}{3}$$

This value being substituted for x in equation (1) gives an imaginary value for y, showing that there is no point in the curve where the tangent is parallel to the axis of abscissas.

MULTIPLE POINTS.

(**155**) A multiple point is one in which two or more branches of a curve intersect each other. At such a point the curve will always have as many tangents as there are branches, and hence $\frac{dy}{dx}$ must have the same number of values for that point.

Let us take the equation

$$y=b\pm(x-a)\sqrt{x-c} \text{ where } a>c \tag{1}$$

then by differentiating we have

$$\frac{dy}{dx}=\pm\sqrt{x-c}\pm\frac{x-a}{2\sqrt{x-c}}$$

For $x=a$ and $x=c$ in equation (1) we have $y=b$; hence H and H′ (Fig. 72) corresponding to these values of x and y are points in the curve. For all values of $x<c$ that of y is imaginary; hence there is no part of the curve between H and the axis of ordinates. For every value of $x>c$, except $x=a$, y has two values, one greater, and the other as much less than b Hence the curve is symmetrical about HH′. For $x=c$ we have $\frac{dy}{dx}=\infty$, hence the tangent at H is perpendicular to the axis of abscissas. For $x=a$ we have two values of $\frac{dy}{dx}$ equal to each other with contrary signs, namely, $\sqrt{x-c}$ and $-\sqrt{x-c}$. Hence at H′ there are two tangents making supplementary angles with the axis of abscissas, so that the two branches of the curve cross each other at that point in directions symmetrical with HH′. If we make $\frac{dy}{dx}=0$ we have

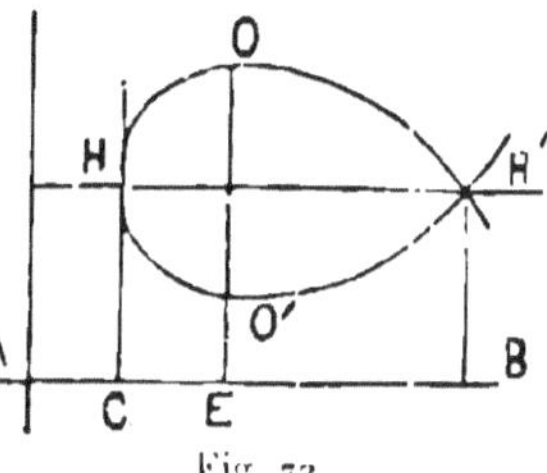

Fig. 72.

$$x=\frac{a+2c}{3}$$

which shows that at the point corresponding with the ordinate at E where AE equals one-third of (2AC+AB), the tangent is parallel to the axis of abscissas.

(156) We will close the discussion of algebraic curves with that of the equation

$$ay^2-x^3+(b-c)x^2+bcx=0$$

Solving this equation with reference to y we have

$$y=\pm\sqrt{\frac{x(x-b)(x+c)}{a}} \qquad (1)$$

and

$$\frac{dy}{dx}=\pm\frac{3x^2-2x(b-c)-bc}{2\sqrt{ax(x-b)(x+c)}}$$

If in equation (1) we make $x=0$, $x=b$, or $x=-c$, we have in every case $y=0$. Hence there are three points, H, A and H′ (Fig. 73), where the curves meet the axis of abscissas.

Every negative value of $x>c$ gives an imaginary value for y, hence the curve has no point on the negative side of H, since $AH=c$. Every negative value of $x<c$ will give two equal values for y with opposite signs; hence from H to A the curve is symmetrical about the axis of abscissas. Every positive value for $x<b$ gives an imaginary value for y; hence no part of the curve lies between A and H′. Every positive value for $x>b$ gives two equal values for y with contrary signs. Hence on the positive side of H′ the curve is symmetrical about the axis of abscissas, and the entire curve consists of two parts having no connection with each other by a common point.

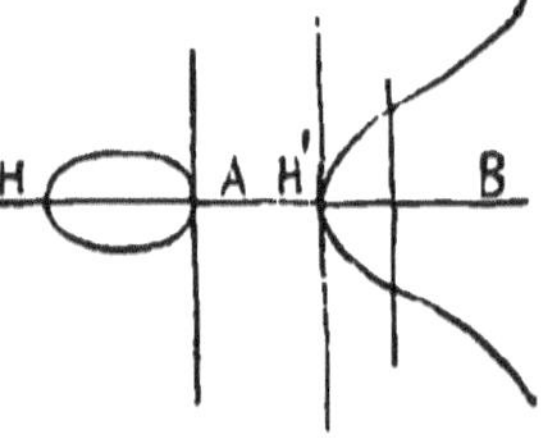

Fig. 73.

Each of the values of x that reduce y to zero also reduce $\frac{dy}{dx}$ to infinity; hence at the points H, A and H′ the tangent is perpendicular to the axis of abscissas, and one of these tangents is the axis of ordinates.

If we solve the equation

$$3x^2-2x(b-c)-bc=0$$

we shall have

$$x=\frac{b-c\pm\sqrt{3bc+(b-c)}}{3}$$

but $\sqrt{3bc+(b-c)^2}<b+c$; hence if we take the positive

value of the radical part the result will be less than $\frac{b-c+b+c}{3}$, that is, less than $\frac{2}{3}b$, hence it will give no point of the curve. If we take the negative value, the result will be numerically less than $-\frac{2}{3}c$; hence there will be two points where the tangent will be parallel to the axis of abscissas, corresponding to the point on that axis where

$$x=\frac{b-c-\sqrt{3bc+(b-c)^2}}{3}.$$

If $c=0$ the equation becomes

$$ay^2=x^3-bx^2$$

in which case the oval HA is contracted into a conjugate point at A as in Art. 154.

If $b=0$ the equation becomes

$$ay^2=x^3+cx^2$$

or

$$y=\pm\sqrt{\frac{x^3+cx^2}{a}}$$

and

$$\frac{dy}{dx}=\pm\frac{3x^2+2cx}{2\sqrt{ax^2(x+c)}}$$

In this case the curve takes the form in Fig. 74. There are two equal values for y with opposite signs for every value of x on the positive side of H where $x=-c$. At that point $\frac{dy}{dx}=\infty$, and the tangent is perpendicular to the axis of abscissas. If we make $\frac{dy}{dx}=0$ we have $x=0$ and $x=-\frac{2c}{3}$; hence the tangent at A and at T and T′ where $x=-\frac{2c}{3}$ are parallel to the axis of abscissas.

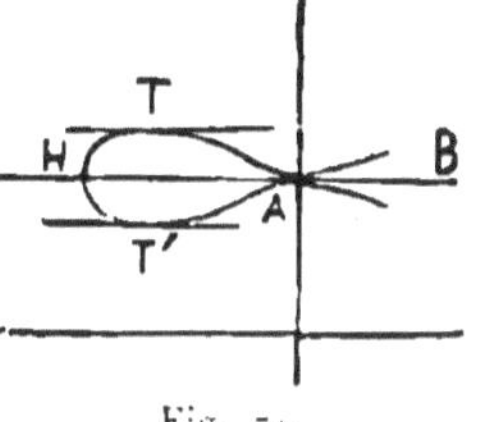

Fig. 74.

If we make both b and c equal to zero we have

$$ay^2 = x^3$$

whence

$$y = \pm\sqrt{\frac{x^3}{a}}$$

and

$$\frac{dy}{dx} = \pm\frac{3}{2}\sqrt{\frac{x}{a}}$$

In this case the curve assumes the form in Fig. 75. There is no negative value for x, and all positive values of x give two equal values for y with contrary signs. At the origin we have $\frac{dy}{dx} = 0$, and hence the axis of abscissas is tangent to both branches of the curve which is a cusp of the first species

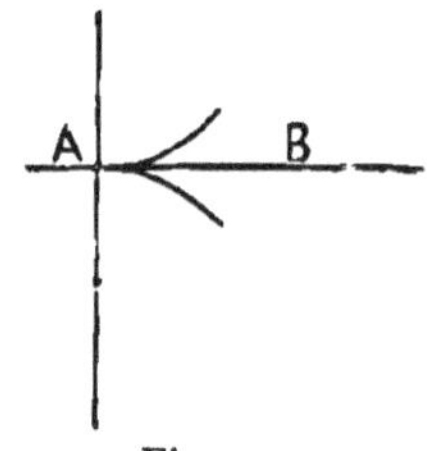

Fig. 75.

PART II.

INTEGRAL CALCULUS.

INTEGRAL CALCULUS.

SECTION I.

PRINCIPLES OF INTEGRATION.

(157) The problem of the differential calculus is to obtain the differential or rate of change in a function arising from that of the variable, or variables, which enter into it. The corresponding problem of the integral calculus is to pass from a given differential of a function to the function itself.

The first of these operations can always be performed directly by rules founded on philosophical principles. The second can only be performed by empirical rules founded on actual experiment. We cannot proceed *directly* from the differential to the function, but, as it were, backwards; that is, we show that a function is the integral of a given differential by showing that the latter would be produced by differentiating the former. Thus we know that x^2 is the integral of $2xdx$, because $2xdx$ has been shown to be the differential of x^2. Hence the rules for integration are merely the rules for differentiation inverted.

While rules have been obtained for differentiating every algebraic function, it by no means follows that every differential can be integrated. The number of simple algebraic functions is very small, and each one has its specific form of differential. Should any function be complicated, it can be analyzed and differentiated in detail, applying only the rules for simple forms. But before a differential can be integrated, it must be reduced to one of the forms arising from differentiating a simple function; and this can be done in comparatively few of the infinite number of forms that differentials may assume. The transformations available for this purpose form one of the chief subjects that demand the attention of the student of the integral calculus. The difficulty of integration is very much increased when the differential is a function of two independent variables, for the rate of change in such a function can give but little indication, generally, what the function is.

There is still another difficulty in obtaining the exact integral of any given differential. We have seen that the constant terms in any function disappear when it is differentiated, and, of course, when we come to integrate an isolated differential expression, we cannot know what constants, if any, should belong to it. In such a case, then, we pay no attention to the question of constants. If, however, the function should occur in an *equation* we can generally find from the conditions expressed by it what value would belong to the constant. Until this is done we indicate by adding the symbol C to the integral that a constant is to be supplied if needed to render the integral *definite*. Until then it is said to be *indefinite*.

The notation indicating the integral of any differential is the letter *s* elongated, thus $\int xdx$ would be read "the integral of xdx." This notation was originally adopted by Leibnitz to indicate the *sum* of the infinitely small differentials or differences of which he supposed the function to be made up, and is still retained as a matter of convenience even by those who reject its original meaning, as employed in the system of Leibnitz.

The following rules for integration are derived from those for differentiating; being in fact the same rules inverted.

(158) *If the differential have a constant coefficient it may be placed without the sign of integration.*

For we have seen (Art. 10) that the differential of a variable having a constant coefficient is equal to the constant multiplied by the differential of the variable; that is to say, the coefficient of the variable will also be the coefficient of its differential; hence the coefficient of the differential will also be the coefficient of its integral, that is, of the variable; and may be placed outside the sign of integration.

Thus

$$d(ax)=adx \text{ hence } \int adx=a\int dx=ax$$

(159) *The integral of a differential function, consisting of any number of terms connected together by the signs plus and minus, is equal to the algebraic sum of the integrals of the terms taken separately.*

For we found (Art. 9) that the differential of a polynomial is found by differentiating each term separately, hence to return from the differential to the polynomial, which is the integral, we must integrate each term separately. Thus

$$d(x+y-z)=dx+dy-dz$$

hence

$$\int(dx+dy-dz)=\int dx+\int dy-\int dz=x+y-z$$

(160) *The integral of a monomial differential consisting of a variable, multiplied by the differential of the variable is equal to the variable raised to a power with an exponent increased by one, and divided by the increased exponent and the differential of the variable.*

We have in (Art. 15) the rule for obtaining the differential of the power of a variable. In other words we have given the steps by which we pass from the power to its differential; and hence to pass back from the differential to its integral, that is, the power, we must retrace each step. Thus in the first case we diminish the exponent by one; in the latter we increase it by one. In the former we multiply by the differential of the variable; in the latter we divide by it. In the former we multiply by the exponent before reducing it; in the latter we divide by the exponent after increasing it. Thus

$$\int nx^{n-1}dx = x^n$$

because

$$d(x^n) = nx^{n-1}dx$$

(161) If the function consist of the power of a polynomial multiplied by its differential, the same rule will apply. Thus let the differential be

$$(ax+x^2)^n(a+2x)dx = (ax+x^2)^n d(ax+x^2)$$

make

$$ax+x^2 = u$$

then

$$(ax+x^2)^n (a+2x)dx = u^n du$$

and

$$\int u^n du = \frac{u^{n+1}}{n+1} = \frac{(ax+x^2)^{n+1}}{n+1}$$

EXAMPLES.

Ex. 1. What is the integral of $\frac{x^2dx}{3}$? *Ans.* $\frac{x^3}{9}$

Ex. 2. What is the integral of $x^{\frac{1}{3}}dx$? *Ans.* $\frac{3x^{\frac{4}{3}}}{4}$

Ex. 3. What is the integral of $\frac{dx}{\sqrt{x}}$? *Ans.* $2\sqrt{x}$

Ex. 4. What is the integral of $\frac{dx}{x^3}$? *Ans.* $-\frac{1}{2x^2}$

Ex. 5. What is the integral of $ax^2dx+\frac{dx}{2\sqrt{x}}$?

Ans. $\frac{ax^3}{3}+\sqrt{x}$

(162) If the exponent of the variable in the case provided for in Art. 160 should be -1, the rule will not apply. For by this rule

$$\int x^{-1}dx=\frac{x^0}{0}=\frac{1}{0}=\infty$$

and this arises from the fact that a differential with such an exponent can never occur under the rule given in Art. 15; for then the variable must have been x^0, a constant quantity, that cannot be differentiated. Such differentials, however, do frequently occur, but the rule for their integration must be drawn from a different source. We have found (Art. 38) that the differential of log. $x=\frac{dx}{x}=x^{-1}dx$, and hence a differential of this kind must be integrated by the rule derived from that given for differentiating logarithms. That is to say, the integral of any fraction, in which the numerator is the differential of the denominator, is the Naperian logarithm of the denominator.

EXAMPLES.

Ex. 1. What is the integral of $\frac{adx}{x}$? *Ans.* a log. x

Ex. 2. What is the integral of $\frac{2bxdx}{a+bx^2}$?

Ans. log. $(a+bx^2)$

Ex. 3. What is the integral of $\frac{adx}{bx}$? *Ans.* $\frac{a}{b}$ log. x

Ex. 4. What is the integral of $\frac{ax^2dx}{x^3}$? *Ans.* a log. x

Ex. 5. What is the integral of $\frac{dx}{a+x}$? *Ans.* log. $(a+x)$

(163) *If the differential be in the form of a polynomial, raised to a power denoted by a positive integral exponent and multiplied by the differential of the variable, the integral may be found by expanding the power and multiplying each term by the differential of the variable. We may then integrate the terms separately.* Thus let us take the expression

$$(a+bx)^2dx$$

Expanding the binomial and multiplying each term by dx we have

$$a^2dx+2abxdx+b^2x^2dx$$

which may be integrated as in Art. 158.

EXAMPLES.

Ex. 1. What is the integral of $(5+7x^2)^2dx$?

Ans. $25x+\frac{70}{3}x^3+\frac{49}{5}x^5$

Ex. 2. What is the integral of $(a+3x^2)^3dx$?

Ans. $a^3x+3a^2x^3+\frac{27}{5}ax^5+\frac{27}{7}x^7$

(164) *If a binomial differential be of such a form that the exponent of the variable without the parenthesis is one less than that of the variable within, the integral will be found by increasing the exponent of the binomial by one and dividing it by the new exponent into the exponent of the variable within into its coefficient.*

For suppose the differential to be

$$(a+bx^n)^m x^{n-1}dx$$

make

$$a+bx^n=p$$

then

$$dp=nbx^{n-1}dx$$

and

$$x^{n-1}dx=\frac{dp}{nb}$$

from which

$$(a+bx^n)^m x^{n-1}dx=\frac{p^m dp}{nb}$$

of which the integral is

$$\frac{p^{m+1}}{(m+1)nb}=\frac{(a+bx^n)^{m+1}}{(m+1)nb}$$

hence the rule.

EXAMPLES.

Ex. 1. What is the integral of $(a+bx^2)^{\frac{1}{2}}mxdx$?

Ans. $\frac{m}{3b}(a+bx^2)^{\frac{3}{2}}$

Ex. 2. What is the integral of $(a^2+x^2)^{-\frac{1}{2}}xdx$?

Ans. $(a^2+x^2)^{\frac{1}{2}}$

Ex. 3. What is the integral of $(a+bx^2)^{\frac{3}{2}}cxdx$?

Ans. $\frac{c(a+bx^2)^{\frac{5}{2}}}{5b}$

(**165**) Every rational fraction, which is the differential of a function of x, may be put under the form

$$\frac{Ax^m+Bx^{m-1}+Cx^{m-2}+\ldots\ldots Dx+E}{Fx^n+Gx^{n-1}+Hx^{n-2}+\ldots\ldots Kx+L}dx$$

in which the greatest exponent of the variable in the denominator exceeds by one or more the greatest exponent in the numerator. For if it is equal or less, a division may be made, until the exponent of the remainder would become less than that of the divisor, and this remainder would become the numerator of the fractional part of the quotient; the other part, consisting of entire terms, would be integrated as in Art. 159. Hence we need only to consider the method of integrating the fractional part of the quotient, or rather any fractions of the form already given.

(**166**) For this purpose we resolve the denominator into factors of the first degree as

$$(x-a)(x-b)(x-c)(x-d) \text{ etc.}$$

and place the fraction under the form

$$\left(\frac{A}{x-a}+\frac{B}{x-b}+\frac{C}{x-c}+\frac{D}{x-d}+\text{ etc.}\right)dx$$

in which A, B, C, D, etc., are constants, whose values are determined by reducing all the fractions to a common denominator, and placing the sum of the numerators equal to the original numerator (the denominators being identical). Since this equality of the numerators must exist, independent of any particular value of x, the coefficients of the like powers of x must be respectively equal to each other; and this will furnish enough equations to determine the values of the constants. Substituting these values the fractions may then be integrated separately.

(167) For example let us take the fraction

$$\frac{2adx}{x^2-a^2} \qquad (1)$$

which by decomposing the denominator may be put into the form

$$\frac{2adx}{(x+a)(x-a)}$$

which we transform into

$$\left(\frac{A}{x+a}+\frac{B}{x-a}\right)dx \qquad (2)$$

which being reduced to a common denominator becomes

$$\frac{Ax-Aa+Bx+Ba}{(x-a)(x+a)}dx$$

Making this last numerator equal to that of (1) we have

$$2a=Ax-Aa+Bx+Ba$$

or

$$(A+B)x+(B-A-2)a=0$$

from which we obtain

$$A+B=0$$

and

$$B-A-2=0$$

whence

$$A=-1 \text{ and } B=1$$

Substituting these values of A and B in (2) we have

$$\frac{2adx}{x^2-a^2}=\frac{dx}{x-a}-\frac{dx}{x+a}$$

and by integration

$$\int \frac{2adx}{x^2-a^2}=\int \frac{dx}{x-a}-\int \frac{dx}{x+a}=\log.(x-a)-\log.(x+a)$$

(**168**) Let us next take the fraction

$$\frac{a^3+bx^2}{a^2x-x^3}dx$$

in which the factors of the denominators are x and (a^2-x^2) or $x(a+x)(a-x)$. If we make

$$\frac{a^3+bx^2}{x(a+x)(a-x)}=\frac{A}{x}+\frac{B}{a-x}+\frac{C}{a+x} \qquad (1)$$

and reduce the second member of the equation to a common denominator we have

$$\frac{a^3+bx^2}{a^3x-x^3}=\frac{Aa^2-Ax^2+Bax+Bx^2+Cax-Cx^2}{x(a-x)(a+x)}$$

and placing the coefficients of the like powers of x in the numerators equal we have

$$B-A-C=b$$
$$Ba+Ca=0$$
$$Aa^2=a^3$$

The last of these equations gives

$$A=a$$

which reduces the first to

$$B-C=a+b$$

and this combined with the second gives

$$B=\frac{a+b}{2} \text{ and } C=-\frac{a+b}{2}$$

Substituting these values of A, B and C in equation (1) we have

$$\frac{a^3+bx^2}{a^3x-x^3}dx=\frac{adx}{x}+\frac{a+b}{2(a-x)}dx-\frac{a+b}{2(a+x)}dx$$

and by integration

$$\int \frac{a^3+bx^2}{a^2x-x^3}dx=a\log.x-\frac{a+b}{2}\log.(a-x)-\frac{a+b}{2}\log.(a+x)$$

which may be reduced to

$$a \log. x-(a+b) \log. (a^2-x^2)$$

NOTE.—The second term of the integral must be negative; for since $d(a-x)$ is $-dx$, we shall have $d(\log. (a-x)=-\frac{dx}{a-x}$ and hence $\frac{dx}{a-x}$ must be the differential of $-\log. (a-x)$.

(169) Let us now take the fraction

$$\frac{xdx}{x^2+4ax-b^2}$$

To find the factors of the denominator we must make it equal to zero, and solve the equation which gives

$$x=-2a\pm\sqrt{4a^2+b^2}$$

and hence the factors of the denominator will be

$$x+2a+\sqrt{4a^2+b^2}$$

and

$$x+2a-\sqrt{4a^2+b^2}$$

To simplify the expression we will represent the constant part of each factor by E and F and we shall have

$$x^2+4ax-b^2=(x+\text{E})(x+\text{F})$$

and we may make

$$\frac{x}{x^2+4ax-b^2}=\frac{\text{A}}{x+\text{E}}+\frac{\text{B}}{x+\text{F}}=\frac{\text{A}x+\text{AF}+\text{B}x+\text{BE}}{(x+\text{E})(x+\text{F})}$$

making the numerators equal we have

$$\text{A}x+\text{AF}+\text{B}x+\text{BE}=x$$

whence

$$\text{A}+\text{B}=1$$

and

$$\text{AF}+\text{BE}=0$$

from which

$$\text{A}=\frac{\text{E}}{\text{E}-\text{F}} \text{ and } \text{B}=-\frac{\text{F}}{\text{E}-\text{F}}$$

Substituting these values of A and B we have

$$\int\frac{xdx}{x^2+4ax-b^2}=\frac{\text{E}}{\text{E}-\text{F}}\int\frac{dx}{x+\text{E}}-\frac{\text{F}}{\text{E}-\text{F}}\int\frac{dx}{x+\text{F}}$$

which becomes by integrating

$$\frac{\mathrm{E}}{\mathrm{E}-\mathrm{F}}\log.(x+\mathrm{E})-\frac{\mathrm{F}}{\mathrm{E}-\mathrm{F}}\log.(x+\mathrm{F})$$

or by substituting the values of E and F

$$\left.\begin{array}{l}\displaystyle\int\frac{xdx}{x^2+4ax-b^2}=\frac{2a+\sqrt{4a^2+b^2}}{2\sqrt{4a^2+b^2}}\log.\left(x+2a+\sqrt{4a^2+b^2}\right)\\ \displaystyle\qquad-\frac{2a-\sqrt{4a^2+b^2}}{2\sqrt{4a^2+b^2}}\log.\left(x+2a-\sqrt{4a^2+b^2}\right)\end{array}\right\}$$

(170) In all these cases the factors of the denominator are unequal. If a part or all of them are equal the rule will not apply. For suppose we have

$$\frac{\mathrm{P}x^4+\mathrm{Q}x^3+\mathrm{R}x^2+\mathrm{S}x+\mathrm{T}}{(x-a)(x-b)(x-c)(x-d)(x-e)}$$

which we make

$$\frac{\mathrm{A}}{x-a}+\frac{\mathrm{B}}{x-b}+\frac{\mathrm{C}}{x-c}+\frac{\mathrm{D}}{x-d}+\frac{\mathrm{E}}{x-e}$$

if some of these factors are equal, say $a=b=c$, we should have

$$\frac{\mathrm{P}x^4+\text{etc.}}{(x-a)^3(x-d)(x-e)}=\frac{\mathrm{A}+\mathrm{B}+\mathrm{C}}{x-a}+\frac{\mathrm{D}}{x-d}+\frac{\mathrm{E}}{x-e}$$

Thus in reducing the second member to a common denominator, A+B+C would have to be considered as a single constant A′, and the three constants A′, D and E would not be sufficient to establish the five equations of condition which are required in making equal the coefficients of the like powers of x. In order to avoid this difficulty we decompose the original fraction and make

$$\frac{\mathrm{P}x^4+\mathrm{Q}x^3+\text{ etc.}}{(x-a)^3(x-d)(x-e)}=\frac{\mathrm{A}+\mathrm{B}x+\mathrm{C}x^2}{(x-a)^3}+\frac{\mathrm{D}}{x-d}+\frac{\mathrm{E}}{x-e}$$

which contains the necessary number of constants, and at the same time, when reduced to a common denominator, will produce a numerator containing x to the fourth power; thus giving a sufficient number of equations between the coefficients of the like powers of x.

In the meantime the expression

$$\frac{A+Bx+Cx^2}{(x-a)^3}$$

may be put into the form

$$\frac{A'}{(x-a)^3}+\frac{B'}{(x-a)^2}+\frac{C'}{x-a}$$

in which A′, B′, C′ are determinate constants. For let

$$x-a=u \text{ then } x=u+a$$

and

$$\frac{A+Bx+Cx^2}{(x-a)^3}=\frac{A+Ba+Ca^2+Bu+2Cau+Cu^2}{u^3}$$

$$=\frac{A+Ba+Ca^2}{u^3}+\frac{B+2Ca}{u^2}+\frac{C}{u}$$

and replacing the value of u we have

$$\frac{A+Bx+Cx^2}{(x-a)^3}=\frac{A+Ba+Ca^2}{(x-a)^3}+\frac{B+2Ca}{(x-a)^2}+\frac{C}{x-a}$$

and since these numerators are constant we may represent them by A′, B′, C′, which gives

$$\frac{A+Bx+Cx^2}{(x-a)^3}=\frac{A'}{(x-a)^3}+\frac{B'}{(x-a)^2}+\frac{C'}{x-a}$$

which is the required form.

As this demonstration may be applied to an expression containing any power of x, we make the proposition a general one, that

$$\left\{\begin{array}{l}\dfrac{Px^{m-1}-Qx^{m-2}+\ldots\ldots Rx+S}{(x-a)^m}= \\ \dfrac{A}{(x-a)^m}+\dfrac{A'}{(x-a)^{m-1}}+\dfrac{A''}{(x-a)^{m-2}}+\text{ etc.}\end{array}\right\}$$

Hence to integrate the expression

$$\frac{Px^4+Qx^3+\text{etc.}}{(x-a)^3(x-d)(x-c)}dx$$

we write

$$\left(\frac{A}{(x-a)^3}+\frac{A'}{(x-a)^2}+\frac{A''}{x-a}+\frac{D}{x-d}+\frac{E}{x-c}\right)dx$$

and reduce these fractions to a common denominator and find the values of A, A′, A″, D, E, in the manner already stated. We shall then have to find the integrals of the following expressions,

$$\frac{E}{x-e}dx,\ \frac{D}{x-d}dx,\ \frac{A''}{x-a}dx,\ \frac{A'}{(x-a)^2}dx,\ \frac{A}{(x-a)^3}dx,$$

the three first we can integrate by the rule for logarithms and the others as follows.

Since dx is the differential of $x-a$ we will make $x-a=z$; then we have

$$\int\frac{Adx}{(x-a)^3}=\int\frac{Adz}{z^3}=\int Az^{-3}\,dz=-\frac{A}{2z^2}=-\frac{A}{2(x-a)^2}$$

and

$$\int\frac{A'dx}{(x-a)^2}=\int\frac{A'dz}{z^2}=-\frac{A'}{z}=-\frac{A'}{x-a}$$

Hence

$$\left\{\begin{array}{l}\int\frac{Px^4+Qx^3+\text{etc.}}{(x-a)^3(x-d)(x-e)}dx=-\frac{A}{2(x-a)^2}-\frac{A'}{x-a}\\ +A''\log.(x-a)+D\log.(x-d)+E\log.(x-e)\end{array}\right\}$$

(**171**) Let us take for example

$$\frac{x^2dx}{x^3-ax^2-a^2x+a^3}$$

the denominator of this fraction may be resolved into the factors

$$(x^2-a^2)(x-a)=(x-a)(x+a)(x-a)$$

or

$$(x-a)^2(x+a)$$

Making then

$$\frac{x^2}{(x-a)^2(x+a)}=\frac{A}{(x-a)^2}+\frac{A'}{x-a}+\frac{B}{x+a}\qquad(1)$$

and reducing the second member to a common denominator, we have

$$\frac{x^2}{(x-a)^2(x+a)}=\frac{A(x+a)+A'(x^2-a^2)+B(x-a)^2}{(x-a)^2(x+a)}$$

Developing the numerator and making the coefficients of the same powers of x equal, we have

$$B + C = 1, \quad A - 2aC = 0, \quad aA - a^2B + a^2C = 0; \qquad (1)$$

from which we obtain

$$A = \tfrac{1}{2}a, \quad B = \tfrac{3}{4}, \quad C = \tfrac{1}{4};$$

hence

$$\frac{x^2dx}{(x-a)^2(x+a)} = \frac{a\,dx}{2(x-a)^2} + \frac{3dx}{4(x-a)} + \frac{dx}{4(x+a)},$$

and

$$\int\frac{x^2dx}{(x-a)^2(x+a)} = -\frac{a}{2(x-a)} + \tfrac{3}{4}\log(x-a) + \tfrac{1}{4}\log(x+a).$$

172. If all the factors of the denominator are equal the expression will be of the form

$$\frac{x^{m-1}dx}{(x-a)^n},$$

and we may integrate this more directly. For let $x-a=z$, then $dx=dz$ and $x=z+a$, hence

$$\frac{x^{m-1}dx}{(x-a)^n} = \frac{(z+a)^{m-1}dz}{z^n}.$$

Expanding $(z+a)^{m-1}$ by the binomial theorem we have

$$\frac{(z+a)^{m-1}}{z^n}dz = \frac{z^{m-1}dz}{z^n} + \frac{(m-1)az^{m-2}dz}{z^n}$$

$$+ \frac{(m-1)(m-2)a^2z^{m-3}dz}{2z^n} + \text{etc.},$$

each of which terms may be integrated separately as in Art. [illegible]

Let us take for example

$$\frac{x^2dx}{(x-a)^3}.$$

173. When two differential functions are equal to each other it does not necessarily follow that their integrals are equal; but if they are not equal, their difference will be a constant quantity for all values of the variable. In other words, if two functions have the same rate of change they will either be equal to each other constantly, or else their difference will be constantly the same. Thus the ages of any two persons increase at the same rate, and they will be therefore of the same age, or else the difference of their ages will always be the same. Two persons traveling in the same direction at the same rate will either be constantly together, or else there will be constantly the same distance between them.

Hence in integrating the members of a differential equation, it becomes an important part of the problem to ascertain if there be any difference between the integrals, and if so, what it is.

To do this we add the undetermined constant C to one member of the integral equation, which shall represent the difference if any. This is called the indefinite integral.

Then, since the difference is constant for all values of the variable, we assign to the variable in one member of the equation some value which will correspond to a known value of the other entire member, and thence obtain the value of C. That value having been substituted in place of C in the integral equation will satisfy it for every value of the variable since it does for one value.

To illustrate these principles, let us take the triangle ABC (Fig. [illegible]). The differential of the surface of this triangle is $axdx$ ([illegible]), a being the tangent of the angle CAB, and A the origin of coordinates. Hence the equation

$$ds = axdx$$

and integrating each member, and adding C to the second, we have

$$S=\frac{ax^2}{2}+C \qquad (1)$$

Now to determine the value of C we give to x a value corresponding to a known value of S. But we know that at the origin in A where $x=o$, we have also $S=o$, and by substituting these values in equation (1) we have

$$o=o+C \text{ hence } C=o$$

and

$$S=\frac{ax^2}{2}$$

is the definite integral.

If now we wish to know the value of any specific part of the triangle, such as ADD′, we make $x=x'=AD$, and we have

$$S=\frac{ax'^2}{2}=\frac{xy}{2}=\frac{AD\times DD'}{2}$$

This is the specific integral.

(**174**) We are not bound, however, to make the value of S commence at the origin where $x=o$. We may if we choose estimate it from any line as DD′. In this case (making $x=AD=x'$) we should have

$$o=\frac{ax'^2}{2}+C$$

whence

$$C=-\frac{ax'^2}{2}=-a\frac{\overline{AD}^2}{2}$$

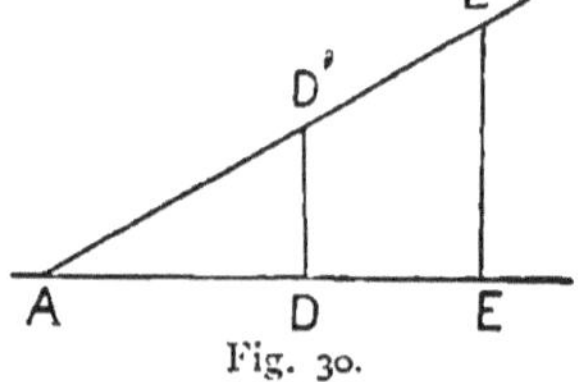

Fig. 30.

and substituting this value in equation (1) we have

$$S=\frac{ax^2}{2}-a\frac{\overline{AD}^2}{2}$$

This is again the definite integral. For any portion of the triangle estimated from DD′ we give the corresponding value of x, say $x=AE=x''$, which gives

$$S=\frac{a}{2}(x''^2-x'^2)=a\frac{\overline{AE}^2-\overline{AD}^2}{2}$$

for the value of the area DD′E′E.

(**175**) There is another method of disposing of the indeterminate constant, which consists in giving to the variable two definite values, and then subtracting one integral from the other. This is called integrating between limits. Thus in the case last noted, if we make x successively equal to $x'=AD$ and $x''=AE$, we shall have

$$S'=\frac{ax'^2}{2}+C \text{ and } S''=\frac{ax''^2}{2}+C$$

and subtracting the first equation from the second we have

$$S'-S''=\frac{a}{2}(x''^2-x'^2)=a\frac{\overline{AE}^2-\overline{AD}^2}{2}$$

the constant C having disappeared in the subtraction.

The notation for this kind of integration consists in placing the two values of the variable at the extremities of the sign of integration; thus

$$\int_{x'}^{x''} axdx$$

indicates that the integral is to be taken between the two values of x represented by x'' and x'; the subtractive one being at the lower extremity of the sign; and the integral would be

$$\frac{ax''^2}{2}-\frac{ax'^2}{2}$$

When the integral is to be taken for any particular value of x, as x', it would be written

$$\int_{x=x'} axdx$$

which indicates that the integral is to be taken where $x=x'$.

EXAMPLES.

Ex. 1. Integrate $2xdx$ between the values of $x=a$ and $x=b$.

Ans. b^2-a^2

Ex. 2. What is the integral of $\int_a^b 3x^2dx$. *Ans.* b^3-a^3

Ex. 3. Integrate $\int_a^b \frac{\pi}{2}x^2dx$. *Ans.* $\frac{\pi}{6}(b^3-a^3)$

Ex. 4. Integrate $\int_a^b 2(c+x)dx$. *Ans.* $b^2+2c(b-a)-a^2$

Ex. 5. Integrate $\int_a^b 3(c+nx^2)^2 2nxdx$.

Ans. $(c+nb^2)^3-(c+na^2)^3$

Ex. 6. Integrate $\int_a^b \frac{dx}{c+x}$. *Ans.* $\log. \frac{c+b}{c+a}$

(176) INTEGRATION BY SERIES.

If it be required to integrate a differential of the form $F(x)dx$ in which $F(x)$ can be developed into a series, the approximate integral may often be found by (Art. 164), and if the series is rapidly converging, its true value may be nearly reached. Let the differential be

$$\frac{dx}{1+x^2}=(1+x^2)^{-1}dx$$

developing by the binomial theorem we have

$$(1+x^2)^{-1}=1-x^2+x^4-x^6+\text{etc.}$$

and multiplying by dx and integrating we have

$$\int\frac{dx}{1+x^2}=x-\frac{x^3}{3}+\frac{x^5}{5}-\frac{x^7}{7}+\text{ etc.}$$

EXAMPLES.

Ex. 1. What is the integral of $\frac{dx}{1+x}$? *Ans.*

Ex. 2. What is the integral of $\frac{dx}{a-x}$? *Ans.*

Ex. 3. What is the integral of $\frac{dx}{(a-x)^2}$? *Ans.*

Ex. 4. What is the integral of $\frac{dx}{\sqrt{1-x^2}}$? *Ans.*

(177) INTEGRATION OF DIFFERENTIALS OF CIRCULAR ARCS.

We have seen (Art. 47) that if u designate the sine of an arc, then the differential of an arc will be

$$\frac{du}{\sqrt{1-u^2}}$$

hence the integral of the function of the form

$$\frac{dx}{\sqrt{1-x^2}}$$

will be an arc of which x is the sine.

(**178**) If the expression is of the form

$$\frac{dx}{\sqrt{a^2-x^2}}$$

we may make $x=av$ then

$$x^2=a^2v^2 \text{ and } a^2-x^2=a^2-a^2v^2=a^2(1-v^2)$$

and

$$dx=adv$$

whence

$$\frac{dx}{\sqrt{a^2-x^2}}=\frac{adv}{a\sqrt{1-v^2}}=\frac{dv}{\sqrt{1-v^2}}$$

and

$$\int\frac{dx}{\sqrt{a^2-x^2}}=\int\frac{dv}{\sqrt{1-v^2}}$$

which is an arc of which $v=\frac{x}{a}$ is the sine.

(**179**) If x represent the cosine of an arc, then the differential of the arc (Art. 47) will be

$$-\frac{dx}{\sqrt{1-x^2}}$$

hence the integral of the form

$$-\frac{dx}{\sqrt{1-x^2}}$$

will be an arc of which x is the cosine.

If the expression is of the form

$$-\frac{dx}{\sqrt{a^2-x^2}}$$

it may be integrated as in (Art. 178)

(**180**) If y represent the tangent of an arc then (Art. 47) the differential of the arc will be

$$\frac{dy}{1+y^2}$$

hence the integral of a function of the form

$$\frac{dx}{1+x^2}$$

will be an arc of which x is the tangent.

(**181**) If the expression is of the form

$$\frac{dx}{a^2+x^2}$$

we may make $x=av$, whence

$$dx=adv \text{ and } a^2+x^2=a^2+a^2v^2=a^2(1+v^2)$$

whence

$$\int\frac{dx}{a^2+x^2}=\int\frac{adv}{a^2(1+v^2)}=\frac{1}{a}\int\frac{dv}{1+v^2}$$

which is equal to $\frac{1}{a}$ into an arc of which $v=\frac{x}{a}$ is the tangent.

(**182**) If we represent the versed sine of an arc by z we have (Art. 47) for the differential of the arc

$$\frac{dz}{\sqrt{2z-z^2}}$$

hence the integral of a function of the form

$$\frac{dx}{\sqrt{2x-x^2}}$$

will be an arc of which x is the versed sine.

(183) If the expression be of the form

$$\frac{dx}{\sqrt{2ax-x^2}}$$

we may assume $x=av$, whence

$dx=adv$ and $2ax-x^2$ becomes $2a^2v-a^2v^2$

or

$$a^2(2v-v^2)$$

whence

$$\int\frac{dx}{\sqrt{2ax-x^2}}=\int\frac{adv}{a\sqrt{2v-v^2}}=\int\frac{dv}{\sqrt{2v-v^2}}$$

which is an arc of which $v=\frac{x}{a}$ is the versed sine.

SECTION II.

INTEGRATION OF BINOMIAL DIFFERENTIALS.

(**184**) The general expression for a binomial differential may be reduced to the form

$$x^{m-1}(a+bx^n)^{\frac{p}{q}}dx \qquad (1)$$

in which m and n are whole numbers, n is positive and x enters but one term of the binomial.

For if m and n are fractional we may substitute another variable with an exponent equal to the least common multiple of the denominators of the given exponents, which will then be reducible to whole numbers.

If, for example, we have

$$x^{\frac{1}{3}}(a+bx^{\frac{1}{2}})^{\frac{p}{q}}dx$$

we make $x=z^6$, then $dx=6z^5dz$, and we have by substitution

$$6z^7(a+bz^3)^{\frac{p}{q}}dz$$

If n is negative we can make $x=\frac{1}{z}$, and the expression would become

$$x^{m-1}(a+bx^{-n})^{\frac{p}{q}}dx=-z^{-m+1}(a+bz^n)^{\frac{p}{q}}dz$$

in which the exponent of z within the parenthesis is positive.

If the expression is of the form

$$x^{m-1}(ax^r+bx^n)^{\frac{p}{q}}dx$$

we may divide the terms within the parenthesis by x^r, and multiply the parenthesis by it, thus

$$x^{m-1}[x^r(a+bx^{n-r})]^{\frac{p}{q}}dx$$

or

$$x^{m-1+\frac{pr}{q}}(a+bx^{n-r})^{\frac{p}{q}}dx$$

thus we may secure the three stated conditions.

(185) If $\frac{p}{q}$ is a whole number and positive, the binomial may be expanded into a finite number of terms and integrated by Art. (163). If it is entire and negative the function becomes a rational fraction.

EXAMPLES.

Ex. 1. Integrate the expression

$$x^2(a+bx^3)^2dx$$

Expanding the binomial we have

$$a^2x^2dx+2abx^5dx+b^2x^8dx$$

and integrating each term separately we obtain for the integral of the binomial differential

$$\int x^2(a+bx^3)^2dx=\frac{a^2x^3}{3}+\frac{abx^6}{3}+\frac{b^2x^9}{9}$$

Ex. 2. Integrate the expression

$$x^3(a+bx^2)^3dx$$

$$Ans.\quad \frac{a^3x^4}{4}+\frac{3a^2bx^6}{6}+\frac{3ab^2x^8}{8}+\frac{b^3x^{10}}{10}$$

Ex. 3. Integrate the expression

$$x^4(a+bx^3)^3dx$$

$$Ans.\quad \frac{a^3x^5}{5}+\frac{3a^2bx^8}{8}+\frac{3ab^2x^{11}}{11}+\frac{b^3x^{14}}{14}$$

Ex. 4. Integrate the expression

$$x^5(a+b^2x^4)^3dx$$

Ans. $\dfrac{a^3x^6}{6}+\dfrac{3a^2b^2x^{10}}{10}+\dfrac{3ab^4x^{14}}{14}+\dfrac{b^6x^{18}}{18}$

(186) *Every binomial differential may be integrated when the exponent of the variable without the parenthesis, increased by one, is exactly divisible by the exponent of the variable within.*

To effect this we substitute for the binomial within the parenthesis, a new variable having an exponent equal to the denominator of the exponent of the parenthesis; thus in the expression

$$x^{m-1}(a+bx^n)^{\frac{p}{q}}dx \qquad (1)$$

we make

$$a+bx^n=z^q \qquad (2)$$

then

$$(a+bx^n)^{\frac{p}{q}}=z^p \qquad (3)$$

From equation (2) we have

$$x=\left(\frac{z^q-a}{b}\right)^{\frac{1}{n}}$$

and raising both members of the equation to the *m*th power we have

$$x^m=\left(\frac{z^q-a}{b}\right)^{\frac{m}{n}}$$

Differentiating and dividing by *m* we have

$$x^{m-1}dx=\frac{q}{nb}\left(\frac{z^q-a}{b}\right)^{\frac{m}{n}-1}z^{q-1}dz \qquad (4)$$

and multiplying together equations (3) and (4) we have

$$x^{m-1}(a+bx^n)^{\frac{p}{q}}dx=\frac{q}{nb}z^{p+q-1}\left(\frac{z^q-a}{b}\right)^{\frac{m}{n}-1}dz$$

If now $\frac{m}{n}$ is an entire positive number, this expression may be integrated by raising $\frac{z^q-a}{b}$ to a power consisting of a limited number of terms, and each term can be integrated separately.

If $\frac{m}{n}$ be negative we may by formula D (Art. 211) increase the exponent until it become positive.

EXAMPLES.

(187) Integrate the expression

$$x^3(a+bx^2)^{\frac{3}{2}}dx$$

Assume

$$a+bx^2=z^2$$

then

$$(a+bx^2)^{\frac{3}{2}}=z^3 \qquad (1)$$

$$x^2=\frac{z^2-a}{b} \qquad (2)$$

an'

$$xdx=\frac{zdz}{b} \qquad (3)$$

Multiplying (1), (2), (3), together we have

$$x^3(a+bx^2)^{\frac{3}{2}}dx=z^4\frac{z^2-a}{b^2}dz$$

of which the integral is

$$\frac{z^7}{7b^2}-\frac{az^5}{5b^2}=\frac{(a+bx^2)^{\frac{7}{2}}}{7b^2}-\frac{a(a+bx^2)^{\frac{5}{2}}}{5b^2}$$

(188) Integrate the expression

$$x^5(a+bx^2)^{\frac{1}{2}}dx$$

Make

$$a+bx^2=z^2$$

then

$$(a+bx^2)^{\frac{1}{2}}=z \qquad (1)$$

$$x^2=\frac{z^2-a}{b} \tag{2}$$

and

$$xdx=\frac{zdz}{b} \tag{3}$$

Squaring (2) and multiplying by (1) and (3) we have

$$x^5(a+bx^2)^{\frac{1}{2}}dx=\left(\frac{z^2-a}{b}\right)^2\frac{z^2dz}{b}=\frac{z^6dz-2az^4dz+a^2z^2dz}{b^3}$$

of which the integral is

$$\frac{z^7}{7b^3}-\frac{2az^5}{5b^3}+\frac{a^2z^3}{3b^3}$$

and restoring the value of z we have

$$\int x^5(a+bx^2)^{\frac{1}{2}}dx=\frac{(a+bx^2)^{\frac{7}{2}}}{7b^3}-\frac{2a(a+bx^2)^{\frac{5}{2}}}{5b^3}+\frac{a^2(a+bx^2)^{\frac{3}{2}}}{3b^3}$$

(189) Integrate the expression

$$x^5(a+bx^2)^{\frac{2}{3}}dx$$

Make

$$a+bx^2=z^3$$

then

$$(a+bx^2)^{\frac{2}{3}}=z^2 \tag{1}$$

also

$$x^2=\frac{z^3-a}{b} \tag{2}$$

and

$$2xdx=\frac{3z^2dz}{b} \tag{3}$$

Multiplying the square of (2) by (1) and (3) we have

$$x^5(a+bx^2)^{\frac{2}{3}}dx=\frac{3z^2}{2b}\left(\frac{z^3-a}{b}\right)^2z^2dz$$

$$=\frac{3z^4}{2b^3}(z^6-2z^3a+a^2)dz$$

$$=\frac{3z^{10}dz}{2b^3}-\frac{3z^7adz}{b^3}+\frac{3a^2z^4dz}{2b^3}$$

which being integrated is

$$\frac{3z^{11}}{22b^3}-\frac{3az^8}{8b^3}+\frac{3a^2z^5}{10b^3}$$

Substituting the value of z we have

$$\int x^5(a+bx^2)^{\frac{2}{3}}dx=\frac{3(a+bx^2)^{\frac{11}{3}}}{22b^3}-\frac{3a(a+bx^2)^{\frac{8}{3}}}{8b^3}+\frac{3a^2(a+bx^2)^{\frac{5}{3}}}{10b^3}$$

(**190**) Integrate the expression

$$x^3(a+x^2)^{-\frac{1}{2}}dx$$

Make

$$a+x^2=z^2$$

then

$$x^2=z^2-a \qquad (1)$$

$$(a+x^2)^{-\frac{1}{2}}=z^{-1} \qquad (2)$$

$$xdx=zdz \qquad (3)$$

Multiplying together (1), (2), (3), we have

$$x^3dx(a+x^2)^{-\frac{1}{2}}=(z^2-a)z^{-1}zdz=(z^2-a)dz$$

and integrating we have

$$\frac{z^3}{3}-az=\frac{(a+x^2)^{\frac{3}{2}}}{3}-a(a+x^2)^{\frac{1}{2}}$$

(**191**) Integrate the expression

$$x^5(a^2+x^2)^{-1}dx$$

Make

$$a^2+x^2=z$$

then

$$x^2=z-a^2 \qquad (1)$$

$$2xdx=dz \qquad (2)$$

and

$$(a^2+x^2)^{-1}=z^{-1} \qquad (3)$$

Multiplying together (2), (3) and the square of (1) we have

$$x^5(a^2+x^2)^{-1}dx=\frac{(z-a^2)^2z^{-1}dz}{2}$$

which being expanded becomes

$$\frac{z^2-2a^2z+a^4}{2}z^{-1}dz=\frac{zdz}{2}-\frac{2a^2dz}{2}+\frac{a^4dz}{2z}$$

Integrating and substituting the value of z we have

$$\int x^5(a^2+x^2)^{-1}dx=\frac{(a^2+x^2)^2}{4}-a^2(a^2+x^2)+\frac{a^4}{2}\log.(a^2+x^2)$$

(**192**) Another condition under which a binomial differential may be integrated is as follows:

Put the expression $x^{m-1}(a+bx^n)^{\frac{p}{q}}dx$ into the following form

$$x^{m-1}\Big[\Big(\frac{a}{x^n}+b\Big)x^n\Big]^{\frac{p}{q}}dx$$

or

$$x^{m-1}x^{\frac{np}{q}}\Big(\frac{a}{x^n}+b\Big)^{\frac{p}{q}}dx=x^{m+\frac{np}{q}-1}(ax^{-n}+b)^{\frac{p}{q}}dx$$

By (Art. 186) this expression is integrable when

$$\frac{m+\frac{np}{q}}{n}=\frac{m}{n}+\frac{p}{q}$$

is a whole number, hence

A binomial may be integrated when the exponent of the variable without the parenthesis, increased by one, divided by the exponent of the variable within the parenthesis, and added to the exponent of the parenthesis is a whole number.

EXAMPLES.

(**193**) Integrate the expression

$$a(1+x^2)^{-\frac{3}{2}}dx$$

Make

$$v^2x^2=1+x^2$$

then

$$(1+x^2)^{-\frac{3}{2}}=v^{-3}x^{-3} \qquad (1)$$

also

$$x^2 = \frac{1}{v^2 - 1}$$

whence

$$dx = \frac{-vdv}{x(v^2 - 1)^2} \qquad (2)$$

and

$$1 = x^4 (v^2 - 1)^2 \qquad (3)$$

Multiplying together (1), (2), (3), we have

$$a(1 + x^2)^{-\frac{3}{2}} dx = -\frac{adv}{v^2}$$

and by integration

$$\int -\frac{adv}{v^2} = \frac{a}{v} = \frac{ax}{\sqrt{1 + x^2}}$$

(194) Integrate the expression

$$dx(a^2 + x^2)^{-\frac{1}{2}} = \frac{dx}{\sqrt{a^2 + x^2}}$$

Make

$$v = x + \sqrt{a^2 + x^2}$$

then

$$dv = dx + \frac{xdx}{\sqrt{a^2 + x^2}} = \frac{x + \sqrt{a^2 + x^2}}{\sqrt{a^2 + x^2}} dx$$

hence

$$\frac{dv}{v} = \frac{dx}{\sqrt{a^2 + x^2}}$$

Representing the integral sought by X_0 we have

$$X_0 = \int \frac{dx}{\sqrt{a^2 + x^2}} = \int \frac{dv}{v} = \log. v = \log. (x + \sqrt{a^2 + x^2})$$

(195) Integrate the expression

$$\frac{x^2 dx}{\sqrt{a^2 + x^2}}$$

Representing the integral by X_2 we have

$$dX_2 = \frac{x^2 dx}{\sqrt{a^2 + x^2}}$$

Make

$$v=(a^2x^2+x^4)^{\frac{1}{2}}$$

then

$$dv=\frac{a^2xdx+2x^3dx}{(a^2x^2+x^4)^{\frac{1}{2}}}=\frac{a^2dx}{\sqrt{a^2+x^2}}+\frac{2x^2dx}{\sqrt{a^2+x^2}}$$

hence

$$dv=a^2dX_0+2dX_2$$

where X_0 has the same value as in (Art. 194). From this we have

$$dX_2=\frac{dv}{2}-\frac{a^2dX_0}{2}$$

or

$$X_2=\frac{v}{2}-\frac{a^2X_0}{2}$$

Replacing the value of X_0 and v we have

$$X_2=\frac{x}{2}(a^2+x^2)^{\frac{1}{2}}-\frac{a^2}{2}\log.(x+\sqrt{a^2+x^2})$$

(196) Integrate the expression

$$x^{-4}(1-x^2)^{-\frac{1}{2}}dx$$

Make

$$v^2x^2=1-x^2$$

then

$$x^{-2}=v^2+1 \text{ and } x^{-4}=(v^2+1)^2 \qquad (1)$$

also

$$x=(v^2+1)^{-\frac{1}{2}} \text{ whence } dx=-(v^2+1)^{-\frac{3}{2}}vdv \qquad (2)$$

and

$$(1-x^2)^{-\frac{1}{2}}=\frac{1}{vx}=\frac{(v^2+1)^{\frac{1}{2}}}{v} \qquad (3)$$

Multiplying together (1), (2), (3), we have

$$x^{-4}(1-x^2)^{-\frac{1}{2}}dx=-(v^2+1)^2(v^2+1)^{-\frac{3}{2}}\frac{(v^2+1)^{\frac{1}{2}}}{v}vdv=-(v^2+1)dv$$

Integrating we have

$$\int x^{-4}(1-x^2)^{-\frac{1}{2}}dx=-\frac{v^3}{3}-v=-\frac{1+2x^2}{3x^3}\sqrt{1-x^2}$$

(197) If a binomial cannot be integrated by any of these methods, there are others to which we may resort. These consist in making such a transformation of the expression that the exponent of the variable without the parenthesis or that of the parenthesis itself may be reduced so as to bring the differential into one of the integrable forms. This is done by separating the differential into parts, one of which shall be an integral quantity, and the other the form to which we desire to reduce the expression. This is called

INTEGRATION BY PARTS.

To effect this we resort to the principle on which the product of two variables is differentiated. We have seen (Art. 11) that

$$d(uv)=udv+vdu$$

hence

$$uv=\int udv+\int vdu$$

or

$$\int udv=uv-\int vdu \qquad (1)$$

If now we can so transform the binomial differential, that while it is represented by the first member of the equation (1), it may also be represented in its transformed state by the second member, we see that the integral may be made to depend on that part represented by vdu, and *that* may be made to assume in certain cases the form of an integrable differential.

The two general methods of doing this are, either to make the part represented by du to contain the variable without the parenthesis with an exponent diminished by that of the variable within the parenthesis; or else to contain the parenthesis itself with an exponent diminished by one. In all other respects this part is to be identical with the given differential binomial.

The following is the first of these methods.

For convenience we represent the exponent of the parenthesis by p, which is supposed to represent a fraction; and substitute m for $m-1$; and we have the general form

$$x^m(a+bx^n)^p dx$$

in which m and n are whole numbers.

Make

$$v=(a+bx^n)^s$$

in which s may have any required value. Differentiating we have

$$dv=bnsx^{n-1}(a+bx^n)^{s-1}dx$$

If we now assume

$$x^m(a+bx^n)^p dx=udv$$

we have

$$u=\frac{x^m(a+bx^n)^p dx}{bnsx^{n-1}(a+bx^n)^{s-1}dx}$$

or

$$u=\frac{x^{m-n+1}(a+bx^n)^{p-s+1}}{bns}$$

which being differentiated gives

$$\left\{\begin{array}{l} du=\dfrac{(m-n+1)x^{m-n}(a+bx^n)^{p-s+1}}{bns}dx \\ \quad +\dfrac{(p-s+1)x^m(a+bx^n)^{p-s}}{s}dx \end{array}\right\}$$

but

$$\left\{\begin{array}{l} (a+bx^n)^{p-s+1}=(a+bx^n)^{p-s}(a+bx^n) \\ \quad =a(a+bx^n)^{p-s}+bx^n(a+bx^n)^{p-s} \end{array}\right\}$$

hence

$$du=\left[\frac{a(m-n+1)x^{m-n}}{bns}+\frac{(m+1+np-ns)x^m}{ns}\right](a+bx^n)^{p-s}dx$$

If now we take the value of s such that

$$m+1+np-ns=0$$

we have

$$s=\frac{m+1}{n}+p$$

and

$$du=\frac{a(m-n+1)x^{m-n}(a+bx^n)^{p-s}}{b(np+m+1)}dx$$

Substituting these values of u, v, du and dv in equation (1) we have

FORMULA **A**

$$\left\{\begin{array}{c}\int x^m(a+bx^n)^p dx= \\ \dfrac{x^{m-n+1}(a+bx^n)^{p+1}-a(m-n+1)\int x^{m-n}(a+bx^n)^p dx}{b(np+m+1)}\end{array}\right\}$$

in which we find the integral of the given differential to depend on the integral of a similar differential in which the exponent of the variable without the parenthesis is diminished by that of the variable within it.

In like manner we should find

$$\int x^{m-n}(a+bx^n)^p dx$$

to depend on

$$\int x^{m-2n}(a+bx^n)^p dx$$

and we may thus continue to diminish the exponent of the variable without the parenthesis as long as it is greater than that of the variable within it.

(**198**) There is frequent occasion to integrate binomials of the form

$$\frac{x^m dx}{\sqrt{a^2-x^2}}$$

Representing its integral by X_m we have

$$X_m=\int\frac{x^m dx}{\sqrt{a^2-x^2}}$$

and substituting in the formula **A** (Art. 197)

-1 for b
2 for n
a^2 for a
$-\frac{1}{2}$ for p

we have

FORMULA a.

$$X_m=\int\frac{x^m dx}{\sqrt{a^2-x^2}}=\frac{(m-1)a^2}{m}\int\frac{x^{m-2}dx}{\sqrt{a^2-x^2}}-\frac{x^{m-1}}{m}\sqrt{a^2-x^2}$$

(**199**) Integrate the expression

$$dX_0=\frac{adx}{\sqrt{a^2-x^2}}$$

We have found (Art. 47) that the differential of the arc of a circle is equal to

$$\frac{\mathrm{R}d\sin.}{\sqrt{\mathrm{R}^2-\sin.^2}}$$

hence we have

X_0=arc of a circle of which a is radius and x is the sine.

(**200**) Integrate the expression

$$X_2=\int\frac{x^2dx}{\sqrt{a^2-x^2}}$$

Substitute in formula a, 2 for m and we have

$$X_2=\frac{a^2}{2}\int\frac{dx}{\sqrt{a^2-x^2}}-\frac{x}{2}\sqrt{a^2-x^2}$$

which is equal to

$$\frac{a}{2}X_0-\frac{x}{2}\sqrt{a^2-x^2}$$

where X_0 has the same value as in (Art. 199).

Similarly by substituting different values for m in formula a we have

$$X_4 = \int \frac{x^4 dx}{\sqrt{a^2 - x^2}} = \frac{3a^2}{4} X_2 - \frac{x^3}{4}\sqrt{a^2 - x^2}$$

$$X_6 = \int \frac{x^6 dx}{\sqrt{a^2 - x^2}} = \frac{5a^2}{6} X_4 - \frac{x^5}{6}\sqrt{a^2 - x^2}$$

$$X_8 = \int \frac{x^8 dx}{\sqrt{a^2 - x^2}} = \frac{7a^2}{8} X_6 - \frac{x^7}{8}\sqrt{a^2 - x^2}$$

in which the values of X_0, X_2, X_4, X_6, X_8 remain the same throughout. Thus formula *a* reduces the integral of a differential of the form

$$\frac{x^m dx}{\sqrt{a^2 - x^2}}$$

to that of one depending on the integrals of differentials of the forms.

$$\frac{x^{m-2} dx}{\sqrt{a^2 - x^2}}, \quad \frac{x^{m-4} dx}{\sqrt{a^2 - x^2}}, \quad \frac{x^{m-6} dx}{\sqrt{a^2 - x^2}} \text{ and so on}$$

until, if m is an even number, we shall after $\frac{m}{2}$ operations find the integral of the given differential to depend on that of a differential of the form

$$\frac{dx}{\sqrt{a^2 - x^2}}$$

which is the differential of the arc of a circle of which $\frac{x}{a}$ is the sine (Art. 177).

(201) By a similar substitution in formula **A** we may find

FORMULA *b*

thus

$$X_m = \int \frac{x^m dx}{\sqrt{a^2 + x^2}} = \frac{x^{m-1}}{m}\sqrt{a^2 + x^2} - \frac{(m-1)a^2}{m}\int \frac{x^{m-2} dx}{\sqrt{a^2 + x^2}}$$

If then we have the expression

$$X_4 = \int \frac{x^4 dx}{\sqrt{a^2 + x^2}}$$

we would make $m=4$ in formula b which would then become

$$X_4=\frac{x^3}{4}\sqrt{a^2+x^2}-\frac{3a^2}{4}\int\frac{x^2dx}{\sqrt{a^2+x^2}}$$

The integral of

$$\frac{x^2dx}{\sqrt{a^2+x^2}}$$

we have found (Art. 195) to be

$$\frac{x}{2}(a^2+x^2)^{\frac{1}{2}}-\frac{a^2}{2}\log.(x+\sqrt{a^2+x^2})$$

hence

$$\int\frac{x^4dx}{\sqrt{a^2+x^2}}=\frac{x^3}{4}\sqrt{a^2+x^2}-\frac{3a^2x}{8}\sqrt{a^2+x^2}+\frac{3a^4}{8}\log.(x+\sqrt{a^2+x^2})$$

(202) The expression

$$X_m=\int\frac{x^m dx}{\sqrt{2ax-x^2}}$$

may be integrated by first reducing it to the given form (Art. 184) and making the proper substitutions in formula **A** (Art. 197).

It may however be integrated by an independent process as follows:

Make

$$v=x^{m-1}\sqrt{2ax-x^2}=(2ax^{2m-1}-x^{2m})^{\frac{1}{2}}$$

and we have by differentiating

$$dv=\frac{a(2m-1)x^{2m-2}dx-mx^{2m-1}dx}{(2ax^{2m-1}-x^{2m})^{\frac{1}{2}}}$$

which becomes by dividing the terms by x^{m-1}

$$dv=\frac{a(2m-1)x^{m-1}dx}{(2ax-x^2)^{\frac{1}{2}}}-\frac{mx^m dx}{(2ax-x^2)^{\frac{1}{2}}}$$

Now this last term is equal to

$$m\,.\,dX_m$$

hence

$$dv=\frac{a(2m-1)x^{m-1}dx}{(2ax-x^2)^{\frac{1}{2}}}-m\,.\,dX_m$$

or by transposition

$$dX_m = \frac{a(2m-1)x^{m-1}dx}{m(2ax-x^2)^{\frac{1}{2}}} - \frac{dv}{m}$$

and by integrating and substituting the value of v we have

FORMULA c

$$\int \frac{x^m dx}{\sqrt{2ax-x^2}} = \frac{a(2m-1)}{m}\int \frac{x^{m-1}dx}{\sqrt{2ax-x^2}} - \frac{x^{m-1}}{m}\sqrt{2ax-x^2}$$

an expression which depends on the integral of

$$\frac{x^{m-1}dx}{\sqrt{2ax-x^2}}$$

in which the exponent of the variable without the parenthesis is diminished by one.

(203) If we take the expression

$$dX_0 = \frac{adx}{\sqrt{2ax-x^2}}$$

we see (Art. 47) that it is the differential of an arc whose radius is a and whose versed sine is x; or, which is the same thing, an arc whose versed sine is $\frac{x}{a}$ and radius 1; hence

$$X_0 = \int \frac{adx}{\sqrt{2ax-x^2}} = \text{ver. sin.}^{-1}\frac{x}{a}$$

(204) If we take the expression

$$X_1 = \int \frac{xdx}{\sqrt{2ax-x^2}}$$

and make m in formula c equal to 1, we shall have

$$X_1 = X_0 - \sqrt{2ax-x^2}$$

in which X_0 has the same value as in (Art. 203).

Similarly

$$X_2 = \int \frac{x^2 dx}{\sqrt{2ax-x^2}} = \frac{3a}{2}X_1 - \frac{x}{2}\sqrt{2ax-x^2}$$

and

$$X_3=\int\frac{x^3dx}{\sqrt{2ax-x^2}}=\frac{5a}{3}X_2-\frac{x^2}{3}\sqrt{2ax-x^2}$$

where X_1, X_2, X_3, have the same value throughout.

Thus formula c reduces the binomial differential.

$$\frac{x^mdx}{\sqrt{2ax-x^2}}$$

to depend successively on the integrals of

$$\frac{x^{m-1}dx}{\sqrt{2ax-x^2}},\quad\frac{x^{m-2}dx}{\sqrt{2ax-x^2}},\quad\frac{x^{m-3}dx}{\sqrt{2ax-x^2}}$$

and finally on

$$\frac{dx}{\sqrt{2ax-x^2}}$$

which represents the differential of an arc whose versed sine is $\frac{x}{a}$ as we have seen.

(**205**) To find the integral of

$$\frac{x^{\frac{3}{2}}dx}{\sqrt{2ax-x^2}}$$

we substitute in formula c, $\frac{3}{2}$ for m which gives

$$\int\frac{x^{\frac{3}{2}}dx}{\sqrt{2ax-x^2}}=\frac{4a}{3}\int\frac{x^{\frac{1}{2}}dx}{\sqrt{2ax-x^2}}-\frac{2x^{\frac{1}{2}}}{3}\sqrt{2ax-x^2}$$

Dividing the terms of the first fraction in the second member of the equation by $x^{\frac{1}{2}}$ we have

$$\frac{x^{\frac{1}{2}}dx}{\sqrt{2ax-x^2}}=\frac{dx}{\sqrt{2a-x}}$$

of which the integral is

$$-2\sqrt{2a-x}$$

hence

(206) The method of diminishing the exponent of the variable without the parenthesis by means of formula **A**, will of course only apply when m is positive. But we may obtain from this another formula which will diminish the exponent when it is negative. To do this we multiply the formula **A** by the denominator and we have

$$\left\{\begin{array}{c} b\,(np+m+1)\int x^m(a+bx^n)^p dx= \\ x^{m-n+1}(a+bx^n)^{p+1}-a(m-n+1)\int x^{m-n}(a+bx^n)^p\,dx \end{array}\right\}$$

or

$$\left\{\begin{array}{c} \int x^{m-n}(a+bx^n)^p\,dx= \\ \dfrac{x^{m-n+1}(a+bx^n)^{p+1}-b(np+m+1)\int x^m(a+bx^n)^p\,dx}{a(m-n+1)} \end{array}\right\}$$

Making $m-n=-m$ we have

FORMULA **B**

$$\left\{\begin{array}{c} \int x^{-m}(a+bx^n)^p\,dx= \\ \dfrac{x^{-m+1}(a+bx^n)^{p+1}-b(np-m+n+1)\int x^{-m+n}(a+bx^n)^p dx}{a(-m+1)} \end{array}\right\}$$

If tn denote the greatest multiple of n contained in m we shall have after $t+1$ reductions the integral of

$$x^{-m}(a+bx^n)^p\,dx$$

to depend on that of

$$x^{-m+(t+1)n}(a+bx^n)^p\,dx$$

and if $-m+(t+1)n=n-1$ we shall have (Art. 164)

$$\int x^{n-1}(a+bx^n)^p\,dx=\frac{(a+bx^n)^{p+1}}{nb(p+1)}$$

but in this case

$$\frac{-m+1}{n}=-t$$

a whole number; and hence the original expression may be integrated as in (Art. 186).

(207) To find the integral of

$$\frac{dx}{x^2(1+x^3)^{\frac{1}{3}}}=x^{-2}(1+x^3)^{-\frac{1}{3}}dx$$

Substitute in formula **B**

2 for m
1 for a
1 for b
3 for n
$-\frac{1}{3}$ for p

and we have

$$\int x^{-2}(1+x^3)^{-\frac{1}{3}}dx=-x^{-1}(1+x^3)^{\frac{2}{3}}+\int x(1+x^3)^{-\frac{1}{3}}dx$$

since $a(-m+1)=-1$ and $b(np-m+n+1)=1$.

(**208**) To find the integral of

$$\frac{dx}{x^2(2-x^2)^{\frac{3}{2}}}=x^{-2}(2-x^2)^{-\frac{3}{2}}ax$$

Substitute in formula **B**

2 for m
2 for a
-1 for b
2 for n
$-\frac{3}{2}$ for p

which gives

$$\int x^{-2}(2-x^2)^{-\frac{3}{2}}dx=-x^{-1}(2-x^2)^{-\frac{1}{2}}-2\int(2-x^2)^{-\frac{3}{2}}dx$$

since $a(-m+1)=-2$ and $b(np-m+n+1)=2$.

(**209**) Besides the method of reducing the exponent of the variable without the parenthesis, we may make the integral to depend on that of another expression of the same form in which the exponent of the parenthesis itself is reduced by one. This is the second general method referred to in (Art. 197).

Let us make $v=x^s$ where s is an exponent to which we may assign any required value. From this we obtain

$$dv=sx^{s-1}dx \qquad (1)$$

If now we assume

$$udv = x^m(a+bx^n)^p dx \tag{2}$$

we shall have by dividing equation (2) by equation (1)

$$u = \frac{x^{m-s+1}}{s}(a+bx^n)^p$$

and

$$du = \frac{m-s+1}{s}x^{m-s}(a+bx^n)^p dx + \frac{bnp}{s}x^{m-s+1}(a+bx^n)^{p-1}x^{n-1}dx$$

but

$$(a+bx^n)^p = (a+bx^n)(a+bx^n)^{p-1}$$

hence

$$du = \frac{a(m-s+1)+b(m-s+1+np)x^n}{s}x^{m-s}(a+bx^n)^{p-1}dx$$

Let the value of s be taken such that

$$m-s+1+np=0$$

or

$$s=m+1+np$$

and we shall have

$$du = \frac{-anpx^{m-s}(a+bx^n)^{p-1}dx}{np+m+1}$$

Substituting these values of u, v, du, dv in formula (1), (Art. 197), we have

FORMULA **C**

$$\int x^m(a+bx^n)^p dx = \frac{x^{m+1}(a+bx^n)^p + anp\int x^m(a+bx^n)^{p-1}dx}{np+m+1}$$

in which the integral of the expression is made to depend on that of one of the same form in which the exponent of the parenthesis is one less than that given.

By a similar process this last may be made to depend on the integral of one whose exponent of the parenthesis is again one less; and so on until the exponent of the parenthesis shall have become less than one.

(210) To integrate the expression

$$dx\sqrt{a^2+x^2}$$

substitute in formula **C**

$$\begin{gathered} 0 \text{ for } m \\ a^2 \text{ for } a \\ 1 \text{ for } b \\ 2 \text{ for } n \\ \tfrac{1}{2} \text{ for } p \end{gathered}$$

and we obtain

$$\int dx\sqrt{a^2+x^2}=\frac{x\sqrt{a^2+x^2}}{2}+\frac{a^2}{2}\int\frac{dx}{\sqrt{a^2+x^2}}$$

but by (Art. 194) we have found

$$\int\frac{dx}{\sqrt{a^2+x^2}}=\log.\,(x+\sqrt{a^2+x^2})$$

hence

$$\int dx\sqrt{a^2+x^2}=\frac{x\sqrt{a^2+x^2}}{2}+\frac{a^2}{2}\log.\,(x+\sqrt{a^2+x^2})$$

in like manner we find

$$\int dx\sqrt{x^2-a^2}=\frac{x\sqrt{x^2-a^2}}{2}-\frac{a^2}{2}\log.\,(x+\sqrt{x^2-a^2})$$

(211) If the exponent of the parenthesis is negative, this formula will, of course, not answer, but we can easily deduce from it one that will effect the object. For this purpose we clear it from fractions, transfer the integral term, and divide by the coefficient of the last term in formula **C** and we have

Formula **D**

$$\left\{\begin{aligned} &\int x^m(a+bx^n)^{p-1}dx \\ &=\frac{-x^{m+1}(a+bx^n)^p+(np+m+1)\int x^m(a+bx^n)^p\,dx}{anp} \end{aligned}\right\}$$

To find the integral of

$$(2-x^2)^{-\frac{3}{2}}dx$$

substitute in formula **D**

$$\begin{array}{r l} 0 & \text{for } m \\ 2 & \text{for } a \\ -1 & \text{for } b \\ 2 & \text{for } n \\ -\frac{3}{2} & \text{for } p-1 \end{array}$$

and we have

$$\int(2-x^2)^{-\frac{3}{2}}dx=\frac{x}{2}(2-x^2)^{-\frac{1}{2}}=\frac{x}{2\sqrt{2-x^2}}$$

To find the integral of

$$\frac{xdx}{(1+x^3)^{\frac{1}{3}}}=x(1+x^3)^{-\frac{1}{3}}dx$$

we substitute in the formula

$$\begin{array}{r l} 1 & \text{for } m \\ 1 & \text{for } a \\ 1 & \text{for } b \\ 3 & \text{for } n \\ -\frac{1}{3} & \text{for } p-1 \end{array}$$

and obtain

$$\int x(1+x^3)^{-\frac{1}{3}}dx=-\frac{x^2}{2}(1+x^3)^{\frac{2}{3}}+2\int x(1+x^3)^{\frac{2}{3}}dx$$

in which $x(1+x^3)^{\frac{2}{3}}dx$ may be developed into a series, and each term integrated separately.

INTEGRATION OF EXPONENTIAL DIFFERENTIALS.

(212) It has been shown (Art. 37), that in the Naperian system, the differential of $a^x=a^x \; log. \; adx$, whence

$$\int a^x dx=\frac{a^x}{log. \; x}$$

By means of this equation we may integrate the general expression $a^x\text{X}dx$, in which X is a function of x. For this purpose we put the expression into the form $X.a^x dx$; and applying the formula (Art. 197)

$$\int u dv = uv - \int v du$$

by making $u = X$, $a^x dx = dv$, and $\frac{a^x}{\log. a} = v$ we have

$$\int X. a^x dx = \frac{Xa^x}{\log. a} - \int \frac{a^x}{\log. a} dX \qquad (1)$$

By differentiating successively the function X and representing the successive differential coefficients by X', X'', X''' . . . . $X^{n'}$ we shall have

$$dX = X'dx,\ dX' = X''dx,\ dX'' = X'''dx, \text{ etc.},$$

whence we have

$$\int \frac{a^x}{\log. a} dX \text{ or } \int \frac{X'}{\log. a} a^x dx = \frac{X'}{(\log. a)^2} a^x - \int \frac{a^x}{(\log. a)^2} dX'$$

and substituting this value for the last term in equation (1)

we have $\int X.a^x dx = \frac{X.a^x}{\log. a} - \frac{X'.a^x}{(\log. a)^2} + \int \frac{a^x}{(\log. a)^2} dX'$

and continuing thus to operate the following series will be developed

$$\int Xa^x dx = a^x \left\{ \frac{X}{\log. a} - \frac{X'}{(\log. a)^2} + \frac{X''}{(\log. a)^3} - \frac{X'''}{(\log. a)^4} \cdots \pm \frac{X^{n'}}{(\log. a)^{n+1}} \right\} \mp \int \frac{a^x d X^{n'}}{(\log. a)^{n+1}}$$

If, in finding the successive differential coefficients X', X'', X'''. . . . $X^{n'}$, the last one of them is constant we shall have $dX^{n'} = 0$ and of course that term of the integral vanishes.

(*b*) Let us take for example $X = x^3$ then we have

$$X' = 3x^2,\ X'' = 2 \,.\, 3x,\ X''' \text{ or } X^{n'} = 2 \,.\, 3$$

and

$$\int x^3 a^x dx = a^x \left(\frac{x^3}{\log. a} - \frac{3x^2}{(\log. a)^2} + \frac{2.3x}{(\log. a)^3} - \frac{6}{(\log. a)^4} \right.$$

We may, also, obtain another development of the function $a^x X dx$ by making $a^x = u$ and $X dx = dv$ in the formula already used. Then representing

$$\int X dx \text{ by } P, \int P dx \text{ by } Q, \int Q dx \text{ by } R, \text{ etc.},$$

and integrating by parts we shall have for $\int u dv = uv - \int v du$

$$\int a^x X dx = a^x P - \int a^x \ log.\ a\ P dx \qquad (2)$$

and

$$\int a^x \ log.\ a\ P dx = a^x \ log.\ a\ Q - \int a^x (log.\ a)^2 Q dx$$

and substituting this value in equation (2) we have

$$\int a^x X dx = a^x P - a^x \ log.\ a\ Q + \int a^x (log.\ a)^2 Q dx$$

and continuing thus to integrate by parts we have in general,

$$\int a^x X dx = a^x (P - Q\ log. a + R\ (log. a)^2 -, \text{etc.}) \ldots \pm \int Z a^x \ (log. a)^n dx$$

If we apply this formula to the case where $X = \frac{1}{x^5}$ we shall find

$$P = -\frac{1}{4x^4},\ Q = -\frac{1}{3 \cdot 4 x^3},\ R = \frac{1}{2 \cdot 3 \cdot 4 x^2},\ Z = \frac{1}{2 \cdot 3 \cdot 4 x}$$

whence

$$\int \frac{a^x dx}{x^5} = a^x \left\{ -\frac{1}{4x^4} - \frac{log.\ a}{3 \cdot 4 \cdot x^3} - \frac{(log.\ a)^2}{2 \cdot 3 \cdot 4 x^2} \right\} - \frac{(log.\ a)^3}{2 \cdot 3 \cdot 4} \int \frac{a^x dx}{x}$$

The integral of $\frac{a^x dx}{x}$ can not be determined exactly by any method yet discovered.

In general we see that whenever tne exponent of x is integral and negative, the integral will finally depend on that of $\frac{a^x dx}{x}$; for in the successive functions P. Q. R. etc., the exponents of x diminish constantly by one; and hence the last function will be

$$\int \frac{A a^x dx}{x} = A \int \frac{a^x dx}{x}$$

A being any constant coefficient. We may obtain an approximate value of the integral of $\frac{A a^x dx}{x}$ by substituting in the expression the development of a^x. If, in Art. 37, we make $v=x$ and $k=log.\ a$ we find

$$a^x=1+x\ log.\ a+\frac{x^2}{2}(log.\ a)^2+\frac{x^3}{2\cdot3}(log.\ a)^3+, \text{etc.},$$

and substituting this value of a^x in the place of that quantity in the expression $\frac{Aa^x dx}{x}$ we have

$$\frac{a^x dx}{x}=\frac{dx}{x}+log.\ a\ dx+\frac{x}{2}(log.\ a)^2 dx+\frac{x^2}{2\cdot3}(log.\ a)^3 dx+, \text{etc.}$$

Integrating each term separately we have

$$\int\frac{a^x dx}{x}=log.\ x+x\ log.\ a+\frac{x^2}{4}(log.\ a)^2+\frac{x^3}{18}(log.\ a)^3+, \text{etc.}$$

If we make $u=x^y$ in the equation $\frac{du}{u}=d\,(log.\ u)$ or rather $du=ud\,(log.\ u)$ we shall have

$$dx^y=x^y d\,(log.\ x^y)$$

so that whenever we can decompose a differential into two factors of which one shall be represented by x^y and the other by $d\,(log.\ x^y)$ the integral will be x^y.

If we make the constant (a) in Eq. (1) equal to the base of the Naperian system of logarithms then, since the logarithm of the base is one, the foregoing results will assume a simpler form.

In the following examples the base of the Naperian system is represented by e.

Ex. 1. Find the integral of $xe^x dx$.

Ans. $e^x(x-1)$.

Ex. 2. Find the integral of $x^2e^x dx$.

Ans. $e^x(x^2-2x+2)$.

Ex. 3. Find the integral of $x^4e^x dx$.

Ans. $e^x(x^4-4x^3+12x^2-24x+24)$.

Ex. 4. Find the integral of $\frac{xdx}{e^x}$

Assume $u=x$, and $dv=e^{-x}dx$ which give

$$du=dx, \text{ and } v=-e^{-x}$$

Hence

$$\int\frac{xdx}{e^x}=-e^{-x}x+\int\frac{dx}{e^x}=-e^{-x}x-e^{-x}.$$

INTEGRATION OF LOGARITHMIC DIFFERENTIALS.

(213) The method of integration by parts may also be applied to integrate the expression

$$Xdx\ (log.\ x)^n$$

for if we represent Xdx by dv, $(log.\ x)^n$ by u and the integral of Xdx by X', we shall have $v=X'$ and

$du=n\,(log.\ x)^{n-1}\dfrac{dx}{x}$; and hence (Art 197)

$$\int Xdx\ (log.\ x)^n=X'\ (log.\ x)^n-n\int\frac{X'}{x}dx\ (log.\ x)^{n-1} \quad (3)$$

So that whenever the integral of Xdx can be found, that of $Xdx(log.\ x)^n$ will depend on the integral of a similar expression in which the exponent of $log.\ x$ is one less than before. Hence if n be entire and positive, the exponent of $log.\ x$ will become zero, and the expression on which the integral will depend will become algebraic.

For example, let us take $X=x^m$, then

$$\int x^m dx=\frac{x^{m+1}}{m+1}=X'$$

and substituting these values in equation (3) we have

$$\int x^m(log.\ x)^n dx=\frac{x^{m+1}}{m+1}(log.\ x)^n-\frac{n}{m+1}\int x^m(log.\ x)^{n-1}dx$$

It will be observed that the term in the second member of the equation to be integrated is of the same form as the original quantity, with $n-1$ instead of n as the exponent of the $log. x$; hence by substituting $n-1$ in place of n we have

$$\int x^m(log.\ x)^{n-1}dx = \frac{x^{m+1}}{m+1}(log.\ x)^{n-1} - \frac{n-1}{m+1}\int x^m(log.\ x)^{n-2}dx$$

and substituting $n-2$ for $n-1$ in this equation we obtain

$$\int x^m(log.\ x)^{n-2}dx = \frac{x^{m+1}}{m+1}(log.\ x)^{n-2} - \frac{n-2}{m+1}\int x^m(log.\ x)^{n-3}dx$$

Continuing thus we shall finally arrive at an expression in which, if n be integral and positive, the last term will be

$$\pm\frac{n(n-1)(n-2)\ \ldots\ 2\ .\ 1}{(m+1)^n}\int x^m(log.\ x)^0 dx$$

and since $(log.\ x)^0 = 1$, the integral of $x^m(log.\ x)^0 dx$ will be

$$\frac{x^{m+1}}{m+1}$$

and the term becomes

$$\pm\frac{n(n-1)(n-2)(n-3)\ \ldots\ 2\ .\ 1}{(m+1)^{n+1}}x^{m+1}$$

so that we shall have

$$\int x^m log.\ x^n dx = \frac{x^{m+1}}{m+1}\left\{(log.\ x)^n - \frac{n(log.\ x)^{n-1}}{m+1}\right.$$

$$\left. + \frac{n(n-1)(log.\ x)^{n-2}}{(m+1)^2} \ldots \pm \frac{n(n-1)(n-2)(n-3)\ldots 2\ .\ 1}{(m+1)^n}\right\} \quad (4)$$

It will be observed that the odd terms are plus and the even ones minus, so that if n be an even number, the last term will be an odd one and plus; if n be an odd number, the last term will be even and minus.

If we make $n=1$ we have from equation (4)

$$\int x^m log.\ x dx = \frac{x^{m+1}}{m+1}log.\ x - \frac{1}{m+1}\cdot\frac{x^{m+1}}{m+1} =$$

$$\frac{x^{m+1}}{m+1}\left\{log.\ x - \frac{1}{m+1}\right\}$$

If we make $m=4$ and $n=1$ we have

$$\int x^4 log.\ x dx=\frac{x^5}{5}\left\{log.\ x-\frac{1}{5}\right\}$$

If we make $n=2$ we have

$$\int x^m(log.\ x)^2 dx=\frac{x^{m+1}}{m+1}\left\{(log.\ x)^2-\frac{2}{m+1}log.\ x+\frac{2}{(m+1)^2}\right\}$$

If we make $m=2$ and $n=2$ we have

$$\int x^2(log.\ x)^2 dx=\frac{x^3}{3}\left\{(log.\ x)^2-\tfrac{2}{3}log.\ x+\tfrac{2}{9}\right\}$$

If we make $m=5$ and $n=3$ we have

$$\int x^5(log.\ x)^3 dx=\frac{x^6}{6}\left\{(log.\ x)^3-\frac{(log.\ x)^2}{2}+\frac{log\ x}{6}-\tfrac{1}{36}\right\}$$

If we make $m=1$ and $n=1$ we have

$$\int x\ log.\ x dx=\frac{x^2}{2}\ (log.\ x-\tfrac{1}{2})$$

If $m=0$ and $n=1$ we have

$$\int log.\ x dx=x(log.\ x-1)$$

If $m=-1$ the second member of the equation (4) becomes infinite. In this case the first member becomes

$$(log.\ x)^n\ \frac{dx}{x}$$

If we make $log.\ x=z$ we have $\frac{dx}{x}=dz$ and

$$\int(log.\ x)^n\ \frac{dx}{x}=\int z^n dz=\frac{z^{n+1}}{n+1}=\frac{(log.\ x)^{n+1}}{n+1}$$

which is true for all values of n except $n=-1$. In this case the expression becomes $\dfrac{dx}{x\ log.\ x}$. If we make $log.\ x=z$ we have $\frac{dx}{x}=dz$ and

$$\int\frac{dx}{x\ log.\ x}=\int\frac{dz}{z}=log.\ z=log.(log.\ x)$$

If n is preceded by the minus sign the method given in equation (4) fails because the exponent of the $log.\ x$, instead of approaching zero, continually, becomes greater. In this case we may put the proposed differential under the form

$$x^{m+1}(log.\ x)^{-n}\frac{dx}{x} \text{ or } x^{m+1}(log.\ x)^{-n}d\ log.\ x$$

assume $u=x^{m+1}$, and $dv=(log.\ x)^{-n}d\ log.\ x$, which give $du=(m+1)x^m dx$ and $v=\frac{(log.\ x)^{-n+1}}{-n+1}$ and we shall have

$$\int x^m log.\ x^{-n}dx=-\frac{x^{m+1}}{(n-1)(log.\ x)^{n-1}}+\frac{m+1}{n-1}\int\frac{x^m dx}{(log.\ x)^{n-1}} \quad (5)$$

If n be a whole number, the repeated application of this formula will at length give for the term to be integrated

$$\frac{x^m dx}{log.\ x}$$

In order to simplify this expression put

$$z=x^{m+1}$$

whence $(m+1)\ log.\ x=log.\ z$, or $log.\ x=\frac{log.\ z}{m+1}$.

Also, by differentiating,

$$(m+1)x^m dx=dz$$

whence

$$x^m dx=\frac{dz}{m+1}$$

Hence

$$\frac{x^m dx}{log.\ x}=\frac{dz}{log.\ z}$$

a differential which can be integrated by a series.

If we make $m=4$ and $n=-2$ we shall have

$$\int\frac{x^4 dx}{(log.\ x)^2}=-\frac{x^5}{log.\ x}+5\int\frac{x^4 dx}{log.\ x}$$

If we make $m=4$ and $n=-3$ we shall have

$$\int\frac{x^4 dx}{(log.\ x)^3}=-\frac{x^5}{2(log.\ x)^2}+\frac{5x^5}{2\ log.\ x}+\frac{25}{2}\int\frac{x^4 dx}{log.\ x}$$

SECTION III.

---

(214) *Application of the Integral Calculus to the Measurement of Geometrical Magnitudes.*

We have seen (Art. 173) that when two differentials are equal, their integrals will also be equal or else have a constant difference. It is upon this principle that the method of measuring geometrical magnitudes by means of the calculus is founded. We obtain the expression for the rate of change in the magnitude, in a function of one variable and its differential. It will follow that the magnitude itself is equal to the integral of the function, or else the difference between them will be constant for all values of the variable.

Thus let M represent any magnitude, and let $F(x)dx$ represent its *rate of increase* while being generated by its element—that is, its *differential*: $F(x)$ being the differential coefficient, and a function of x.

Then we have

$$dM = F(x)dx$$

which is an equation between two differentials, hence the integrals are equal or else differ by a constant quantity. If we represent the integral of $F(x)dx$ by X we shall have

$$M = X + C$$

where C represents the constant difference between the quantities whose rates of change are equal.

The method of disposing of the term C is shown in (Art. 173), and the result will be an expression for the value of

M in terms of one variable. Then assigning to this variable any specific value, we obtain the value of M from the beginning up to that value of the variable; or, by giving to the variable two successive values, the difference of the two resulting expressions will give the value of that portion of M lying between the two values of the variable.

RECTIFICATION OF CURVES.

(215) To rectify a curve is to find what would be its length if it were developed into a straight line; in other words, to find the measure of its length. When its differential can be obtained in an integrable form it is said to be *rectifiable*.

The general expression for the differential of any plane curve whose equation is referred to rectangular axes is (Art. 34)

$$du=\sqrt{dx^2+dy^2}$$

and hence

$$u=\int\sqrt{dx^2+dy^2}+C$$

is the general expression for an indefinite portion of any such curve. In order to obtain the integral of this expression, we must know the relation between x and y which we obtain from the equation of the curve; and by means of it eliminate one of the variables and its differential from the formula; thus producing a differential function involving but one variable and its differential, whose integral, when it can be obtained, will be the length of an indefinite portion of the curve.

(216) *To find the length of a Circular Arc.*

We have in (Art. 47) several expressions for the differential of an arc of a circle, in terms of its trigonometrical functions, which already contain but one variable in each case.

If we select that in which the tangent is the variable, we will represent it by t and the formula becomes

$$du=\frac{dt}{1+t^2}=\frac{1}{1+t^2}dt$$

Developing the fraction we have

$$\frac{1}{1+t^2}=1-t^2+t^4-t^6+t^8-\text{etc.}$$

hence

$$du=\frac{1}{1+t^2}dt=dt-t^2dt+t^4dt-t^6dt+\text{etc.}$$

and integrating each term separately we have

$$\int du=u=t-\frac{t^3}{3}+\frac{t^5}{5}-\frac{t^7}{7}+\frac{t^9}{9}-\text{etc.}$$

or

$$u=t\left(1-\frac{t^2}{3}+\frac{t^4}{5}-\frac{t^6}{7}+\frac{t^8}{9}-\text{etc.}\right)+C$$

But we have found (Art. 180) $C=0$, and if we assume

$$u=30^\circ$$

we shall have

$$t=\sqrt{\tfrac{1}{3}}$$

and substituting this value for t we have

$$u=\sqrt{\tfrac{1}{3}}\left(1-\frac{1}{3\cdot 3}+\frac{1}{5\cdot 3^2}-\frac{1}{7\cdot 3^3}+\frac{1}{9\cdot 3^4}-\text{etc.}\right.$$

which being reduced is equal to 0.523598 nearly, for the length of an arc of 30°; and multiplying by 6 we have the arc of a semi-circle equal to 3.141588 when radius is 1. Hence this is the ratio between the diameter and the entire circumference.

(217) *To find the length of an Arc of a Parabola.*

We have found (Art. 57) that the differential of an arc of a parabola is

$$du=\frac{dy}{p}\sqrt{p^2+y^2}$$

The integral of this (Art. 210) is

$$u=\frac{y}{2p}\sqrt{p^2+y^2}+\frac{p}{2}\log.(y+\sqrt{p^2+y^2})+C$$

If we estimate from the vertex of the curve where $u=0$ and $y=0$ we shall have

$$o=\frac{p}{2}\log.p+C$$

hence

$$C=-\frac{p}{2}\log.p$$

Substituting this value of C we have for the definite integral

$$u=\frac{y}{2p}\sqrt{p^2+y^2}+\frac{p}{2}\log.(y+\sqrt{p^2+y^2})-\frac{p}{2}\log.p$$

or

$$u=\frac{y}{2p}\sqrt{p^2+y^2}+\frac{p}{2}\log.\frac{y+\sqrt{p^2+y^2}}{p}$$

(218) *To find the length of an Arc of an Ellipse.*

We have found (Art. 57) that the differential of an arc of an ellipse is

$$du=\frac{1}{A}dx\sqrt{\frac{A^4-(A^2-B^2)x^2}{A^2-x^2}}.$$

If we take c to represent the distance from the center to the focus of the ellipse we have

$$B^2=A^2-c^2$$

hence

$$du=\frac{1}{A}dx\sqrt{\frac{A^4-c^2x^2}{A^2-x^2}}$$

If now we represent the eccentricity of the ellipse by e we have $c=Ae$, and hence

$$du=\frac{1}{A}dx\sqrt{\frac{A^4-A^2e^2x^2}{A^2-x^2}}$$

or, dividing by A^4 under the radical and multiplying by A^2 without it we have

$$du = A dx \sqrt{\frac{1-\frac{e^2x^2}{A^2}}{A^2-x^2}}$$

Developing $\left(1-\frac{e^2x^2}{A^2}\right)^{\frac{1}{2}}$ by the binomial theorem we have

$$\left(1-\frac{e^2x^2}{A^2}\right)^{\frac{1}{2}} = 1-\frac{e^2x^2}{2A^2}-\frac{e^4x^4}{2\,.\,4\,.\,A^4}-\frac{3e^6x^6}{2\,.\,4\,.\,6\,.\,A^6}-\text{etc.}$$

hence

$$\left\{\begin{array}{l} du=\dfrac{A dx}{\sqrt{A^2-x^2}}-\dfrac{e^2}{2A}\cdot\dfrac{x^2dx}{\sqrt{A^2-x^2}}-\dfrac{e^4}{2\,.\,4A^3}\cdot\dfrac{x^4dx}{\sqrt{A^2-x^2}} \\ \qquad -\dfrac{3e^6}{2\,.\,4\,.\,6A^5}\cdot\dfrac{x^6dx}{\sqrt{A^2-x^2}}-\text{etc.}\end{array}\right\}$$

Making

$$\frac{A dx}{\sqrt{A^2-x^2}}=dX_0,\ \frac{x^2dx}{\sqrt{A^2-x^2}}=dX_2,\ \frac{x^4dx}{\sqrt{A^2-x^2}}=dX_4,\ \text{etc.}$$

we have

$$\int du = X_0-\frac{e^2}{2A}X_2-\frac{e^4}{2\,.\,4A^3}X_4-\frac{3e^6}{2\,.\,4\,.\,6A^5}X_6-\text{etc.} \qquad (1)$$

Now by (Art. 199) $X_0=$ the arc of a circle of which A is the radius and x the sine, and (Art. 200)

$$X_2=\frac{A}{2}X_0-\frac{x}{2}\sqrt{A^2-x^2}$$

also

$$X_4=\frac{3A^2}{4}X_2-\frac{x^3}{4}\sqrt{A^2-x^2}$$

and

$$X_6=\frac{5A^2}{6}X_4-\frac{x^5}{6}\sqrt{A^2-x^2}$$

If we make $x=0$ and estimate from the extremity of the conjugate axis, we have $u=0$ and $C=0$. If we make $x=A$ we shall have $u=$ a quadrant, and since

$$A^2-x^2=0$$

we have

$$X_2=\frac{A}{2}X_0,\ X_4=\frac{3A^2}{4}X_2=\frac{3A^3}{2\cdot 4}X_0,\ X_6=\frac{3\cdot 5A^5}{2\cdot 4\cdot 6}X_0$$

and substituting these values in equation (1) we have

$$u=X_0\left(1-\frac{e^2}{2\cdot 2}-\frac{3e^4}{2\cdot 2\cdot 4\cdot 4}-\frac{3\cdot 3\cdot 5e^6}{2\cdot 2\cdot 4\cdot 4\cdot 6\cdot 6}-\text{etc.}\right)$$

for one-fourth of the circumference of an ellipse; X_0 being one-fourth of the circumference of a circle of which the diameter is equal to the major axis of the ellipse.

Hence the whole circumference is equal to

$$2\pi A\left(1-\frac{e^2}{2\cdot 2}-\frac{3e^4}{2\cdot 2\cdot 4\cdot 4}-\frac{3\cdot 3\cdot 5e^6}{2\cdot 2\cdot 4\cdot 4\cdot 6\cdot 6}-\text{etc.}\right.$$

It will be seen that as the eccentricity diminishes the circumference of the ellipse approaches the value of $2\pi A$, which it reaches when $e=0$, and the curve becomes a circle.

(219) *To find the length of the Arc of a Cycloid.*

We have found (Art. 129) that the differential equation of a cycloid is

$$dx=\frac{y\,dy}{\sqrt{2ry-y^2}}$$

By substituting this value of dx in the formula we have

$$du=\sqrt{dx^2+dy^2}=dy\sqrt{\frac{y^2}{2ry-y^2}+1}=dy\sqrt{\frac{2ry}{2ry-y^2}}.$$

or

$$u=\sqrt{2r}\int dy(2r-y)^{-\frac{1}{2}}=-2\sqrt{2r}(2r-y)^{\frac{1}{2}}=-2\sqrt{2r(2r-y)}+C$$

If we estimate the arc from D′ (Fig. 54) where $y=2r$ we shall have $u=0$ and $C=0$, and hence making $y=FG$

$$u=D'F=-2\sqrt{2r(2r-y)} \qquad (1)$$

We see from the figure that

$$D'E'=2r \text{ and } D'H'==2r-y$$

hence

$$\sqrt{2r(2r-y)}=\sqrt{DE \cdot D'H'}=D'F'$$

so that the arc of a cycloid is equal to twice the corresponding chord of the generating circle.

If we take the arc D'A, the corresponding chord of the generating circle becomes the diameter D'E', and half the arc of the cycloid is equal to twice the diameter of the generating circle, or the entire arc is equal to four times that diameter. Thus, if we make $y=0$ we have

$$u=4r$$

or

$$D'FA=2D'E' \text{ and } AD'B=4D'E'$$

(220) *To find the length of the Arc of a Logarithmic Spiral.*

We have found (Art. 77) that the differential of an arc of a polar curve is

$$du=\sqrt{r^2dv^2+dr^2}$$

and the equation of the logarithmic spiral is

$$v=\text{Log. } r$$

Hence

$$dv=\frac{Mdr}{r} \text{ and } dv^2=\frac{M^2dr^2}{r^2}$$

Substituting this value of dv^2 we have

$$du=\sqrt{M^2dr^2+dr^2}=dr\sqrt{M^2+1}$$

In the Naperian system $M=1$ and

$$du=dr\sqrt{2}$$

hence

$$u=r\sqrt{2}+C$$

If we estimate the arc from the pole where $r=0$ we shall have $C=0$ and

$$u=r\sqrt{2}$$

That is, the length of an arc of a Naperian logarithmic spiral estimated from the pole is equal to the diagonal of a square of which the radius vector is the side.

(221) *To find the length of an Arc of the Spiral of Archimedes.*

The equation of the spiral (Art. 84) is

$$r = av$$

in which $a = \frac{1}{2\pi}$ and $v =$ the arc of the measuring circle whose radius is the value of r after one revolution. Hence

$$du = \sqrt{r^2 dv^2 + dr^2} = a dv \sqrt{1 + v^2}$$

the integral of which may be found in (Art. 210). Substituting 1 for a and v for x, thus

$$u = a\left(\frac{v\sqrt{1+v^2}}{2} + \tfrac{1}{2}\log.(v + \sqrt{1+v^2})\right) + C$$

Estimating the arc from the pole where $v = 0$ we shall have $C = 0$ and

$$u = \frac{1}{4\pi}[v + \sqrt{1+v^2} + \log.(v + \sqrt{1+v^2})]$$

(222) *To find the length of an Arc of a Hyperbolic Spiral.*

The equation of this spiral is (Art. 86)

$$rv = ab \text{ or } r = \frac{ab}{v}$$

Differentiating we have

$$dr = -\frac{ab dv}{v^2}$$

whence

$$du = \sqrt{\frac{a^2 b^2 dv^2}{v^4} + \frac{a^2 b^2}{v^2} dv^2} = \frac{ab dv}{v}\sqrt{\frac{1}{v^2} + 1}$$

and

$$u = ab \int v^{-2} dv \sqrt{v^2 + 1}$$

Integrating by formula B (Art. 206); making in the formula

$$m = 2$$

$$a = 1$$

$$b = 1$$

$$p = \tfrac{1}{2}$$

$$n = 2$$

we have

$$\int v^{-2}dv\sqrt{v^2+1}=-v^{-1}(1+v^2)^{\frac{3}{2}}+2\int dv(1+v^2)^{\frac{1}{2}}$$

or (Art. 221)

$$=-v^{-1}(1+v^2)^{\frac{3}{2}}+2\left[\frac{v\sqrt{1+v^2}}{2}+\tfrac{1}{2}\log.(v+\sqrt{1+v^2})\right]$$

hence

$$u=ab[-v^{-1}(1+v^2)^{\frac{3}{2}}+v\sqrt{1+v^2}+\log.(v+\sqrt{1+v^2})]+C$$

Estimating the arc from the point where $v=0$ we have

$$u=ab(\tfrac{1}{0})^{\frac{3}{2}}=\infty$$

which is as it should be, since from the equation of the curve, when $v=0$ the radius vector is infinite. As v is infinite when $r=0$ we shall have $u=0$ at the same time. Hence the curve is unlimited in but one direction. We may, however, find the length of any intermediate portion by substituting the two corresponding values of v in the integral function and taking the difference of the results.

(223) *Quadrature of Curves.*

The quadrature of a curve is the process of finding the measure of a plain surface bounded wholly, or in part, by a curve.

To find the area of such a surface we must find its differential in a function of one variable, which being integrated will give an expression for an *indefinite* portion of the area, from which any *specific* portion may be obtained by assigning corresponding values to the variable.

(224) *To find the area of a Semi-Parabola.*

We have (Art. 65) for the differential of the surface of a parabola

$$dS=ydx=\sqrt{2p}x^{\frac{1}{2}}dx$$

of which the integral is

$$S=\tfrac{2}{3}\sqrt{2p}\,x^{\frac{3}{2}}=\tfrac{2}{3}x\sqrt{2px}=\tfrac{2}{3}xy+C$$

But when $x=0$ we have $S=0$, and hence $C=0$; so that the surface of a parabola bounded by the curve, the axis and an ordinate is equal to two-thirds of the rectangle described on the ordinate and corresponding abscissa.

(225) *To find the area of any Parabola.*

The general equation of the parabola is

$$y^n=ax$$

from which we obtain

$$dx=\frac{ny^{n-1}dy}{a}$$

hence

$$S=\int ydx=\int\frac{ny^n\,dy}{a}=\frac{ny^{n+1}}{(n+1)a}=\frac{n}{n+1}xy+C$$

If we estimate the curve from the origin where $S=0$ we have $x=0$, and hence $C=0$, and

$$S=\frac{n}{n+1}xy$$

That is, the area of that portion of any parabola, bounded by the curve, the axis and the ordinate, is equal to the rectangle described upon the ordinate and corresponding abscissa, multiplied by the ratio $\frac{n}{n+1}$. If $n=2$, as in the common parabola, we have

$$S=\tfrac{2}{3}xy$$

If $n=\frac{2}{3}$, as in the cubic parabola, we have

$$S=\tfrac{2}{5}xy$$

If $n=1$, the figure becomes a triangle and we have

$$S=\tfrac{1}{2}xy$$

or half the base into the height.

(226) *To find the area of a Circle.*

We have (Art. 63) for the circle

$$ydx=dx\sqrt{R^2-x^2}$$

Making R=1 we have

$$dS=ydx=dx\sqrt{1-x^2}=dx(1-x^2)^{\frac{1}{2}}$$

Developing the binomial and multiplying each term by dx we have

$$dS=dx-\frac{x^2dx}{2}-\frac{x^4dx}{8}-\frac{x^6dx}{16}-\frac{5x^8dx}{128}-\text{etc.}$$

from which we obtain by integrating each term separately

$$S=x-\frac{x^3}{6}-\frac{x^5}{40}-\frac{x^7}{112}-\frac{5x^9}{1152}-\text{etc.}+C$$

Estimating the area from the center where $x=0$ we have S=0, and, therefore, C=0, so that the series expresses the area of any segment between the ordinate at the center where $x=0$ and the ordinate corresponding to any other value of x. Hence if we make $x=1$ we have the area of a quadrant equal to

$$1-\tfrac{1}{6}-\tfrac{1}{40}-\tfrac{1}{112}-\tfrac{5}{1152}-\text{etc.}$$

which by taking enough terms may be reduced to

$$.78539$$

Hence the entire area of the circle will be equal to

$$3.14156$$

equal to π where radius is 1.

(227) We may also find the area of a circle by considering it as being described by the revolution of the radius about the center. In this case the radius of the circle becomes the radius vector and we have (Art. 82)

$$dS=\frac{r^2dv}{2R}$$

where v represents the arc of the measuring circle and R its radius. Integrating the terms of this equation we have, since r is constant,

$$S=\frac{r^2v}{2R}+C$$

Estimating the area from the beginning where S=0 we have $v=0$, and hence C=0, and

$$S=\frac{r^2v}{2R} \tag{1}$$

is the measure of a sector of a circle of which v=the measuring arc. Making $R=r$ we shall have v=the arc of the given circle, and equation (1) becomes

$$S=\frac{rv}{2}$$

that is, the measure of a sector of a circle is half the product of the radius into the arc of the sector; and hence for the entire circle, the area is equal to half the product of the radius into the circumference.

If we make v= the entire circumference we have

$$v=2\pi R$$

and substituting this value in equation (1) we have

$$S=\frac{r^2v}{2R}=\pi r^2$$

that is the area of a circle is equal to the square of the radius multiplied by the ratio between the diameter and the circumference.

(228) *To find the area of an Ellipse.*

We have in the case of an ellipse (Art. 64)

$$dS=ydx=\frac{B}{A}dx\sqrt{A^2-x^2}$$

hence

$$S=\frac{B}{A}\int(A^2-x^2)^{\frac{1}{2}}dx$$

Integrating by formula **C** (Art. 209), and substituting

o for m

A^2 for a

-1 for b

2 for n

$\frac{1}{2}$ for p

we have

$$\int(A^2-x^2)^{\frac{1}{2}}dx=\frac{x(A^2-x^2)^{\frac{1}{2}}}{2}+\frac{A^2}{2}\int(A^2-x^2)^{-\frac{1}{2}}dx$$

but (Art. 178)

$$\int(A^2-x^2)^{-\frac{1}{2}}dx=\int\frac{dx}{\sqrt{A^2-x^2}}=\sin.^{-1}\frac{x}{A}$$

hence

$$S=\frac{B}{2A}x\sqrt{A^2-x^2}+\frac{AB}{2}\sin.^{-1}\frac{x}{A}+C$$

Estimating from the center where $x=0$ we have $S=0$, and hence $C=0$. Making then $x=A$ we have

$$S=\frac{AB}{2}\sin.^{-1}1=\frac{AB}{2}\cdot\frac{\pi}{2}$$

for one-fourth of the area of the ellipse, since the arc whose sine is 1 is equal to one-fourth of the whole circumference; and we have for the area of the entire ellipse

$$S=\pi AB$$

We may also observe that (Art. 63)

$$dx\sqrt{A^2-x^2}$$

is the differential of the area of a circle whose radius is A, hence the area of an ellipse is $\frac{B}{A}\times$ the area of the circumscribing circle which is πA^2; and is, therefore, equal to

$$\frac{B}{A}\cdot\pi A^2=\pi AB$$

If $A=B$ the expression becomes

$$\pi A^2 \text{ or } \pi R^2$$

for the area of a circle.

(229) *To find the area of a Segment of a Hyperbola.*

We have in the case of a hyperbola

$$y=\frac{B}{A}\sqrt{x^2-A^2}$$

whence

$$dS=ydx=\frac{B}{A}dx\sqrt{x^2-A^2}$$

Integrating by formula **C** (Art. 209) we have

$$S=\frac{B}{2A}x\sqrt{x^2-A^2}-\frac{AB}{2}\log.(x+\sqrt{x^2-A^2})+C$$

To find the value of C we make $x=A$ where $S=0$, and we have

$$0=-\frac{AB}{2}\log.A+C$$

hence

$$C=\frac{AB}{2}\log.A$$

and

$$S=\frac{B}{2A}x\sqrt{x^2-A^2}-\frac{AB}{2}\log.\left(\frac{x+\sqrt{x^2-A^2}}{A}\right)$$

which represents a portion of the area between the curve and the ordinate lying on one side of the axis. Hence the area of the entire segment cut off by the double ordinate is

$$\frac{B}{A}x\sqrt{x^2-A^2}-AB\log.\left(\frac{x+\sqrt{x^2-A^2}}{A}\right)$$

but

$$\frac{B}{A}\sqrt{x^2-A^2}=y$$

hence

$$S=xy-AB\log.\left(\frac{x}{A}+\frac{y}{B}\right)=xy-AB\log.\left(\frac{Ay+Bx}{AB}\right)$$

for the value of the area.

(230) We may also find the area of that part of the surface lying between the curve and the asymptotes, by using the equation of the hyperbola referred to its center and asymptotes, which is

$$xy=m$$

but as the asymptotes are not usually at right angles to each other, we must introduce into the expression for the differential of this area, the *sine* of the angle which they make with each other (Art. 58) which we will call v. We shall then have

$$dS=\sin. v . ydx=\sin. v\frac{mdx}{x}$$

and

$$S=\sin. v . m . \log. x+C$$

If we call the abscissa of the vertex 1, and estimate from the corresponding ordinate, we shall have at that point

$$m=1,\ S=0,\ \log. x=0 \text{ and hence } C=0$$

And since sin. v may be considered as the modulus of a system of logarithms, we may make

$$S=M . \log. x=\text{Log. } x$$

That is, the area between the curve and the asymptote, estimated from the ordinate of the vertex, is equal to the logarithm of the abscissa, taken in a system whose modulus is the sine of the angle made by the asymptotes with each other.

(231) *To find the area of a Cycloid.*

We have (Art. 129)

$$dx=\frac{ydy}{\sqrt{2ry-y^2}}$$

hence

$$dS=ydx=\frac{y^2\,dy}{\sqrt{2ry-y^2}}$$

Integrating this by formula c (Art. 202) we have

$$S=\tfrac{3}{2}r . \text{ver. sin.}^{-1}y-\frac{3r+y}{2}\sqrt{2ry-y^2}+C$$

Estimating the integral from A (Fig. 76) where $y=0$, we have

$$S=0 \text{ and hence } C=0$$

and taking the integral where

$$y=2r=DE$$

we have

$$S=\tfrac{3}{2}r . \text{ver. sin.}^{-1}2r=3\frac{\pi r^2}{2}$$

that is, the area ADE is equal to three times the semi-circle DF′E. Hence the entire area of the cycloid is equal to three times the area of the generating circle.

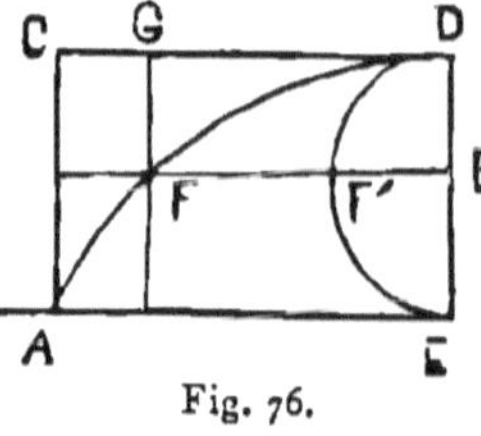

Fig. 76.

(**232**) Another method of obtaining the area of a cycloid is, to consider that portion of the rectangle ACDE which lies outside the curve.

If we make $GF=2r-y=v$ we shall have the differential of the area DCAF equal to vdx or

$$dS'=(2r-y)dx=(2r-y)\frac{ydy}{\sqrt{2ry-y^2}}=dy\sqrt{2ry-y^2}$$

Now if we take the equation of a circle with the origin at the extremity of the diameter we shall have

$$ydx=dx\sqrt{2rx-x^2}$$

which is the differential of the segment of a circle of which x is the abscissa. Hence $dy\sqrt{2ry-y^2}$ is the differential of a circle of which y is the abscissa, that is of the segment F′BE. Hence the two areas ACGF and F′BE have the same rate of change or differential for the same value of y; and since they are both equal to zero when $y=0$, they are equal for every other value of y (Art. 173), and, of course, when $y=2r$. Hence

$$ACD=DF'E=\frac{\pi r^2}{2}$$

But the rectangle $ACDE=\pi r\,.\,2r=2\pi r^2$, hence

$$ADE=ACDE-ACD=\frac{3\pi r^2}{2}$$

as we found in (Art. 231).

(**233**) *To find the area bounded by the coordinate axes and the logarithmic curve.*

We have had (Art. 137) for the logarithmic curve

$$x=\text{Log.}\,y \text{ and } dx=\frac{Mdy}{y}$$

hence

$$dS = ydx = Mdy \text{ and } S = My + C$$

If we estimate the area from AD (Fig. 56) where $y=1$ we have

$$0 = M + C$$

whence

$$C = -M$$

and

$$S = M(y-1)$$

If we make $y=0$ we have

$$S = -M = \text{area ADD}'$$

If $y=2=P'T$ we have

$$S = M = \text{area ADP}'\text{T}$$

So that although the axis of abscissas is an asymptote (Art. 88) to the curve on the negative side, and, therefore, will not meet it within a finite distance, yet the area enclosed between them is limited and equal to ADP′T.

(234) If we take the curve represented by the equation

$$y^2 = \frac{1}{x}$$

to which the axes of coordinates are asymptotes, we shall find a case somewhat similar.

Putting the equation into the form

$$x = \frac{1}{y^2}$$

we have

$$dx = -\frac{2dy}{y^3}$$

and

$$dS = ydx = -\frac{2dy}{y^2}$$

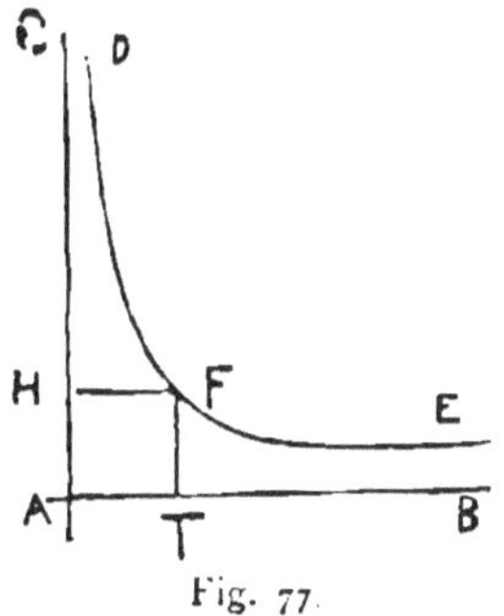

Fig. 77.

hence

$$S = \frac{2}{y} + C$$

If we estimate the area from the line AC where $y=\infty$ we have

$$0=0+C$$

hence

$$C=0 \text{ and } S=\frac{2}{y}$$

If we make $y=1=FT$ we have

$$S=2=ATDC$$

that is, the area ATDC is equal to twice the square AHFT, and is, therefore, finite, although the curve FD does not meet the axis AC at a finite distance.

If we take the area between the limits $y=1$ and $y=0$, we shall have

$$S=\frac{2}{0}-2$$

that is, the area FEBT is infinite, although AB is likewise an asymptote to the curve.

(235) *To find the area described by the radius vector of the Spiral of Archimedes.*

The differential of the area of a polar curve (Art. 82) is

$$\frac{r^2 dv}{2R}$$

v being the arc of the measuring circle and R its radius.

The equation of the Spiral of Archimedes (Art. 84) is

$$r=av$$

hence

$$dr=adv$$

or

$$dv=\frac{dr}{a}$$

Hence, making $R=1$ we have

$$S=\int\frac{r^2 dv}{2}=\int\frac{r^2 dr}{2a}=\frac{r^3}{6a}+C$$

If we make $r=0$ we have

$$S=o \text{ and hence } C=o$$

and since $a=\frac{r}{v}$ we have

$$S=\frac{r^2 v}{6}$$

Making $v=2\pi$ we have for the area described by one revolution of the radius vector

$$S=\frac{\pi r^2}{3}$$

that is, the area described by one revolution of the radius vector is one-third of the area of the circle described with a radius equal to the last value of the radius vector.

If the radius vector make two revolutions we have $v=4\pi$ and

$$S=\frac{2\pi r'^2}{3}$$

where $r'=2r$ and

$$S=\frac{8\pi r^2}{3}$$

But in making two revolutions, the radius vector describes the first part of the area twice. This, therefore, must be subtracted, and we have the area enclosed by the curve and radius vector after two revolutions equal to

$$\tfrac{8}{3}\pi r^2-\tfrac{1}{3}\pi r^2=\tfrac{7}{3}\pi r^2$$

and by subtracting the first again we have the *increased* area described during the second revolution equal to

$$\tfrac{7}{3}\pi r^2-\tfrac{1}{3}\pi r^2=2\pi r^2$$

After m revolutions we have

$$S=\frac{m\pi r'^2}{3}$$

where $r'=mr$, hence

$$S=\frac{m\pi m^2 r^2}{3}=\frac{m^3\pi r^2}{3} \qquad (1)$$

Subtracting from this the area described by the radius vector during $(m-1)$ revolutions we have

$$S' = \frac{\pi r^2}{3}(m^3 - (m-1)^3) \tag{2}$$

Substituting $(m+1)$ in place of m we have

$$S'' = \frac{\pi r^2}{3}((m+1)^3 - m^3) \tag{3}$$

for the area described by the radius vector during the $(m+1)th$ revolution. Taking the difference between equations (2) and (3) we have the *additional* area described by the $(m+1)th$ revolution of the radius vector, that is the area lying between the mth and $(m+1)th$ spires thus

$$S'' - S' = \frac{\pi r^2}{3}((m+1)^3 - 2m^3 + (m-1)^3) = 2m\pi r^2$$

We have found the additional area described by the radius vector during the second revolution equal to $2\pi r^2$, hence the additional area described during the $(m+1)th$ revolution is equal to m times that described by the radius vector during the second revolution. That is, the *increase* of the *additional* areas described by the radius vector during successive revolutions, is *uniform* and equal to twice the area of the circle described with a radius equal to the radius vector after one revolution.

If the area ABP (Fig. 34) be required, that is, the *additional* area corresponding to the arc BC described after the first revolution, we shall have

$$v = 2\pi + \frac{2\pi}{n}$$

and

$$r' = r + \frac{r}{n}$$

and the required area will be

$$S = \frac{\pi + \frac{\pi}{n}}{3}\left(r + \frac{r}{n}\right)^2 - \frac{\pi r^2}{3}$$

or

$$S=\frac{\pi(n+1)}{3n}\cdot\frac{r^2(n+1)^2}{n^2}-\frac{\pi r^2}{3}$$

$$=\frac{\pi r^2}{3n^3}(n+1)^3-\frac{\pi r^2}{3}$$

Developing $(n+1)^3$ we have

$$S=\frac{\pi r^2}{3n^3}(n^3+3n^2+3n+1)-\frac{\pi r^2}{3}$$

or

$$S=\frac{\pi r^2}{3}\left(1+\frac{3}{n}+\frac{3}{n^2}+\frac{1}{n^3}\right)-\frac{\pi r^2}{3}=\frac{\pi r^2}{n}\left(1+\frac{1}{n}+\frac{1}{3n^2}\right)$$

If BCD$=\frac{1}{4}$ circumference $=\frac{2\pi}{4}$, then $n=4$ and

$$ABP=\frac{\pi}{4}r^2(1+\tfrac{1}{4}+\tfrac{1}{48})$$

(236) *To find the area of the surface described by the radius vector of the Hyperbolic Spiral.*

The equation of the hyperbolic spiral (Art. 86) is

$$rv=ab \text{ or } r=\frac{ab}{v}$$

in which a is the radius of the measuring circle and b is the unit of the measuring arc. Hence

$$S=\int\frac{r^2dv}{2a}=\int\frac{ab^2dv}{2v^2}=-\frac{ab^2}{2v}$$

which is infinite when $v=0$, and zero when $v=\infty$. If we make $v=b=$AB (Fig. 35), and $v=\frac{1}{2}b=$As, we shall have

$$S=ab-\frac{ab}{2}=\frac{ab}{2}=\frac{PB\,.\,OR}{2}$$

(237) *To find the area described by the radius vector of a Logarithmic Spiral.*

We have (Art. 87) for this spiral

$$v=\text{Log. } r$$

and

$$dv=\frac{\mathrm{M}dr}{r}$$

Substituting this value in the formula, and making R=1 and M=1, we have

$$\mathrm{S}=\int\frac{r^2dv}{2\mathrm{R}}=\int\frac{rdr}{2}=\frac{r^2}{4}+\mathrm{C}$$

If we estimate from the pole where S=0 we have r=0, and hence C=0, and

$$\mathrm{S}=\frac{r^2}{4}$$

That is, the area described by the radius vector of the Napeaian logarithmic spiral is equal to one-fourth of the square described upon the last value of the radius vector.

(238) *Areas of Surfaces of Revolution.*

We have seen (Art. 66) that the area of a surface of revolution, where the axis of revolution is the axis of abscissas, is

$$\mathrm{S}=\int 2\pi y\sqrt{dx^2+dy^2}$$

the radical part being the differential of an arc of the revolving curve.

To apply this formula to a particular case we must obtain from the equation of the revolving curve, the value of one variable and its differential in terms of the other, so that, when substituted in the formula we may have the differential of the surface in terms of one variable which can then be integrated.

(239) *To find the convex surface of a Cone.*

We have (Art. 67) for the differential of the convex surface of a cone

$$d\mathrm{S}=2\pi axdx\sqrt{a^2+1}$$

in which x is the length of the axis and a is the tangent of the angle made by the revolving element of the cone with the axis of revolution. Integrating we have

$$S=\pi ax^2\sqrt{a^2+1}+C$$

Estimating from the vertex where S=o we have x=o, and hence C=o and

$$S=\pi ax^2\sqrt{a^2+1}$$

But from the equation of the generating line we have

$$y=ax$$

and hence

$$a^2=\frac{y^2}{x^2}$$

and by substitution we have

$$S=\pi yx\sqrt{\frac{y^2+x^2}{x^2}}=\pi y\sqrt{x^2+y^2}$$

or (Fig. 31) making x=AB

$$S=\pi CB\sqrt{AB^2+CB^2}$$

that is, the convex surface of a cone is equal to the circumference of the base multiplied by half the slant height.

(240) For the convex surface of a cylinder we have

$$y=R=\text{radius of the base}$$

hence

$$S=\int 2\pi y\sqrt{dx^2+dy^2}=\int 2\pi R dx=2\pi Rx$$

that is, the convex surface of a cylinder is equal to the circumference of its base into its altitude

(241) In the case of the sphere we have (Art. 68)

$$dS=2\pi R dx$$

hence

$$S=2\pi Rx+C$$

Estimating from the center where x=o we have S=o and hence C=o, and the measure of an indefinite portion of the convex surface of a sphere is

$$S=2\pi Rx$$

the same as that of the circumscribing cylinder having the same altitude.

Making $x=\text{R}$ we have

$$\text{S}=2\pi\text{R}^2$$

for the measure of the convex surface of half the sphere; hence for the entire sphere we have

$$\text{S}=4\pi\text{R}^2$$

or four great circles.

(**242**) *To find the surface of an Ellipsoid described by an ellipse revolving about its major axis.*

Making

$$\sqrt{dx^2+dy^2}=du$$

we have

$$2\pi y\sqrt{dx^2+dy^2}=2\pi y du$$

But we have found (Art. 218)

$$du=\frac{\text{A}dx}{\sqrt{\text{A}^2-x^2}}\left(1-\frac{e^2x^2}{2\text{A}^2}-\frac{e^4x^4}{2\,.\,4\text{A}^4}-\frac{3e^6x^6}{2\,.\,4\,.\,6\text{A}^6}-\text{etc.}\right)$$

hence

$$d\text{S}=\frac{2\pi\text{A}ydx}{\sqrt{\text{A}^2-x^2}}\left(1-\frac{e^2x^2}{2\text{A}^2}-\frac{e^4x^4}{2\,.\,4\,.\,\text{A}^4}-\frac{3e^6x^6}{2\,.\,4\,.\,6\,.\,\text{A}^6}-\text{etc.}\right)$$

But

$$\frac{\text{A}y}{\sqrt{\text{A}^2-x^2}}=\text{B}$$

hence

$$d\text{S}=2\pi\text{B}dx\left(1-\frac{e^2x^2}{2\,.\,\text{A}^2}-\frac{e^4x^4}{2\,.\,4\,.\,\text{A}^4}-\frac{3e^6x^6}{2\,.\,4\,.\,6\,.\,\text{A}^6}-\text{etc.}\right)$$

Integrating each term separately we have

$$\text{S}=2\pi\text{B}x\left(1-\frac{e^2x^2}{2\,.\,3\,.\,\text{A}^2}-\frac{e^4x^4}{2\,.\,4\,.\,5\,.\,\text{A}^4}-\frac{3e^6x^6}{2\,.\,4\,.\,6\,.\,7\,.\,\text{A}^6}-\text{etc}+\text{C}\right.$$

Taking the integral between the limits

$$x=0 \text{ and } x=\text{A}$$

we have for half the surface of the ellipsoid

$$\text{S}=2\pi\text{BA}\left(1-\frac{e^2}{2\,.\,3}-\frac{e^4}{2\,.\,4\,\,5}-\frac{3e^6}{2\,.\,4\,.\,6\,.\,7}-\text{etc.}\right)$$

and multiplying this expression by 2 we have the measure of the entire surface of the ellipsoid.

If we make A=B, then e=o, and we have for the surface of the sphere

$$S=4\pi R^2$$

as before.

(243) *To find the surface of a Paraboloid of revolution.*

We have (Art. 69) in the case of the paraboloid

$$dS=2\pi y\sqrt{\frac{y^2+p^2}{p^2}dy^2}=\frac{2\pi y dy}{p}(y^2+p^2)^{\frac{1}{2}}$$

Integrating according to (Art. 164) we have

$$S=\frac{2\pi}{3p}(y^2+p^2)^{\frac{3}{2}}+C$$

Estimating from the vertex where S=o and y=o we have

$$o=\frac{2\pi}{3p}p^3+C=\frac{2\pi}{3}p^2+C$$

hence

$$C=-\frac{2\pi p^2}{3}$$

and

$$S=\frac{2\pi}{3p}((y^2+p^2)^{\frac{3}{2}}-p^3)$$

(244) *To find the area of the surface described by a Cycloid revolving about its base.*

We have (Art. 129) in the case of the cycloid

$$dx=\frac{ydy}{\sqrt{2ry-y^2}}$$

hence

$$\sqrt{dx^2+dy^2}=\sqrt{\frac{y^2dy^2}{2ry-y^2}+dy^2}=dy\sqrt{\frac{2ry}{2ry-y^2}}$$

and by substitution

$$dS=2\pi y\sqrt{dx^2+dy^2}=2\pi y dy\sqrt{\frac{2ry}{2ry-y^2}}$$

or

$$S=2\pi\sqrt{2r}\int\frac{y^{\frac{3}{2}}dy}{\sqrt{2ry-y^2}}$$

But we have found (Art. 205)

$$\int\frac{y^{\frac{3}{2}}dy}{\sqrt{2ry-y^2}}=-\frac{8r}{3}\sqrt{2r-y}-\frac{2y}{3}\sqrt{2r-y}$$

hence

$$S=2\pi\sqrt{2r}\,(-\frac{8r}{3}\sqrt{2r-y}-\frac{2y}{3}\sqrt{2r-y})+C$$

If we estimate the surface from the plane passing through the middle point of the base we shall have $S=0$ when $y=2r$, hence $C=0$. Then making $y=0$ we have for half the surface required

$$S=2\pi\sqrt{2r}\left(\frac{8r}{3}\sqrt{2r}\right)=\tfrac{32}{3}\pi r^2$$

and for the entire surface twice that quantity. That is, the area of the surface described by the revolution of a cycloid about its base is equal to twenty-one and one-third times that of the generating circle.

(245) *The Cubature of Solids.*

The cubature of a solid is to find the dimensions of an equivalent cube or other known volume.

We have (Art. 71)

$$\pi y^2 dx$$

for the differential of a solid of revolution where y is the ordinate and x the abscissa of the bounding line of the revolving surface which generates the solid; and the axis of abscissas is the axis of revolution. Hence

$$V=\int\pi y^2 dx$$

To apply the formula to any particular solid or volume, we eliminate one of the variables by means of the equation of the bounding curve, thus producing a differential function of one variable which may be integrated.

(246) *To find the volume of a Right Cone.*

Making the vertex the origin we have

$$y=ax$$

for the equation of the bounding line; but a is the tangent of the angle made by this line with the axis of the cone, and is equal to $\frac{b}{h}$, where b is the radius of the base and h the the length of the axis; hence

$$y=\frac{b}{h}x \text{ and } y^2=\frac{b^2}{h^2}x^2$$

whence

$$V=\int\pi y^2dx=\int\pi\frac{b^2}{h^2}x^2dx=\pi\frac{b^2}{h^2}\cdot\frac{x^3}{3}+C$$

Estimating from the origin we have $V=0$, $x=0$, and hence $C=0$, and making $x=h$ we have for the entire cone

$$V=\pi b^2\frac{h}{3}$$

that is, the volume of a cone is equal to one-third of the product of its base by its altitude, or equal to one-third of a cylinder of the same base and altitude.

(247) *To find the volume of a Sphere.*

From the equation of the circle we have

$$y^2=R^2-x^2$$

hence

$$V=\int\pi y^2dx=\int\pi(R^2-x^2)dx=\pi(R^2x-\frac{x^3}{3})+C$$

Estimating from the plane passed through the center, where

$x=0$, we have $S=0$ and $C=0$, and making $x=R$ we have for half the volume of the sphere

$$V=\tfrac{2}{3}\pi R^3$$

and for the entire sphere

$$V=\tfrac{4}{3}\pi R^3$$

Since the surface of the sphere is equal to $4\pi R^2$ we have the volume equal to the surface multiplied by one-third of the radius.

(248) *To find the volume of an Ellipsoid.*

Taking the origin at the extremity of the transverse axis we have for the equation of the bounding line or curve

$$y^2=\frac{B^2}{A^2}(2Ax-x^2)$$

hence

$$V=\int\pi y^2 dx=\pi\int\frac{B^2}{A^2}(2Ax-x^2)dx=\pi\frac{B^2}{A^2}\left(Ax^2-\frac{x^3}{3}\right)+C$$

Estimating from the origin where $x=0$ we have $V=0$, and hence $C=0$; and making $x=2A$ we have

$$V=\pi\frac{B^2}{A^2}(4A^3-\tfrac{8}{3}A^3)=\pi\tfrac{4}{3}B^2A=\tfrac{2}{3}\pi B^2 \,.\, 2A$$

that is, the volume of a prolate ellipsoid is equal to two-thirds of the volume of a cylinder having the minor axis for its diameter, and whose altitude is equal to the major axis.

If the ellipse is made to revolve about its minor axis we should have

$$V'=\pi\tfrac{4}{3}A^2B$$

for the volume of an oblate ellipsoid, and hence

$$V : V' :: \pi\tfrac{4}{3}B^2A : \pi\tfrac{4}{3}A^2B :: B : A$$

that is, the volume of a *prolate* ellipsoid is to that of an *oblate* ellipsoid generated by the same ellipse, as the minor axis is to the major axis.

If $A=B$ we have

$$V=\tfrac{4}{3}\pi R^3$$

as before.

(249) *To find the volume of a Paraboloid.*

In this case we have

$$y^2=2px$$

and

$$V=\int\pi y^2 dx=2\pi p\int x dx=2\pi p\frac{x^2}{2}=\pi p x^2+C$$

Estimating from the vertex where $x=o$ we have $V=o$, and hence $C=o$, and

$$V=\pi p x^2=\frac{\pi y^2 x}{2}$$

or, the volume of a paraboloid is equal to half the volume of the circumscribing cylinder.

(250) *To find the volume of a Solid described by the revolution of a cycloid about its base.*

Since in the case of the cycloid

$$dx=\frac{y dy}{\sqrt{2ry-y^2}}$$

we have

$$V=\int\pi y^2 dx=\pi\int\frac{y^3 dy}{\sqrt{2ry-y^2}}$$

But we have found (Art. 204)

$$\int\frac{y^3 dy}{\sqrt{2ry-y^2}}=\frac{5r}{3}X_2-\frac{y^2}{3}\sqrt{2ry-y^2}$$

$$X_2=\frac{3r}{2}X_1-\frac{y}{2}\sqrt{2ry-y^2}$$

$$X_1=X_0-\sqrt{2ry-y^2}$$

$X_0=$ arc of a circle of which r is radius and y the versed sine.

Integrating between the limits $y=0$ and $y=2r$ we shall have half the volume required; but $y=0$ gives $V=0$ and $C=0$ and $y=2r$ gives

$$X_0=\pi r$$
$$X_1=X_0=\pi r$$
$$X_2=\frac{3r}{2}X_1=\frac{3\pi r^2}{2}$$

and

$$V=\pi\left(\frac{5r}{3}X_2-\frac{y^2}{3}\sqrt{2ry-y^2}\right)=\frac{5\pi^2r^3}{2}$$

hence the entire volume is

$$V=5\pi^2r^3$$

But the volume of the circumscribing cylinder is

$$4\pi r^2 . 2\pi r=8\pi^2r^3$$

Hence the volume of the solid generated by the revolution of a cycloid about its base is five-eighths of that of the circumscribing cylinder.

Appendix.

ANSWERS

TO EXAMPLES FOR PRACTICE.

ARTICLE 10.

Ex. 4.—*Ans.* $dx - bdy$.

Ex. 5.—*Ans.* $(a+b)dx$.

Ex. 6.—*Ans.* $(c-d)dy$.

Ex. 7.—*Ans.* $adx + bdy + cdz$.

Ex 8.—*Ans.* $b^2du + c^2dz$.

Ex. 9.—*Ans.* $a^2bdx + c^2dy$.

Ex. 10.—*Ans.* $a^2dy - b^2dx$.

Ex. 11.—*Ans.* $b(ady - cdx)$.

Ex. 12.—*Ans.* $c^2(bdx + adz)$.

ARTICLE 13.

Ex. 4.—*Ans.* $xdy + ydx - udz - zdu$.

Ex 5.—*Ans.* $3adx - 2xdy - 2ydx$.

Ex. 6.—*Ans.* $2ady + 3du$.

Ex. 7.—*Ans.* $4ab(yzdx + xzdy + xydz)$

Ex. 8.—*Ans.* $bcdu - zdy - ydz$.

Ex. 9.—*Ans.* $4a(xdy + ydx) + cdu$.

Ex. 10.—*Ans.* $-2bdu + cdy$.

ARTICLE 14.

Ex. 3.—*Ans.* $-\frac{adx}{x^2}+\frac{cdy}{y^2}$

Ex. 4.—*Ans.* $\frac{(b-a)dx}{(x-c)^2}$

Ex. 5.—*Ans.* $5dx+\frac{2(c-y)dx+2xdy}{(c-y)^2}$

Ex. 6.—*Ans.* $-3dx-\frac{(u-d)dy-(y-c)du}{(u-d)^2}$

Ex. 7.—*Ans.* $2adx+(y-c)dx+xdy.$

Ex. 8.—*Ans.* $4(u-c)(bdy-dx)+4(by-x)du.$

Ex. 9.—*Ans.*

$$\frac{-(b-y)dx+(a-x)dy}{(b-y)^2}\times\frac{d+u}{v}+\frac{a-x}{b-y}\times\frac{vdu-(d+u)dv}{v^2}$$

Ex. 10. *Ans.* $-(xy-z)dv+(a-v)(xdy+ydx-dz).$

Ex. 11.—*Ans.* $-3(xdy+ydx)(u-c)-3xydu.$

Ex. 12.—*Ans.*

$$(dz-dy)(a-x)-(z-y)dx-\frac{(y-z)dx+(c-x)(dy-dz)}{(y-z)^2}$$

ARTICLE 19.

Ex. 3.—*Ans.* $2axdx-3bx^2dx+dx.$

Ex. 4.—*Ans.* $(c+d)(2ydy-2xdx).$

Ex. 5.—*Ans.* $25x^4dx-2ady.$

Ex. 6.—*Ans.* $(nx^{n-1}-3x^2)dx.$

Ex. 7.—*Ans.* $3ax^2dx-3bdx.$

Ex. 8.—*Ans.* $(x^2+a)dx+2x(x-a)dx.$

Ex. 9.—*Ans.* $2xy^2dx+2x^2ydy-2zdz.$

Ex. 10.—*Ans.* $2ax(x^3+a)dx+3ax^4dx.$

Ex. 11.—*Ans.* $\dfrac{4aydy}{(b-2y^2)^2}$

Ex. 12.—*Ans.* $-\dfrac{xdx}{\sqrt{a^2-x^2}}$

Ex. 13.—*Ans.* $\dfrac{(a+x)dx}{\sqrt{2ax-x^2}}$

Ex. 14.—*Ans.* $\dfrac{xdx}{(1-x^2)^{\frac{3}{2}}}$

Ex. 15.—*Ans.* $\dfrac{dx}{(x+\sqrt{1-x^2})^2\sqrt{1-x^2}}$

Ex. 16.—*Ans.* $\dfrac{3(a+\sqrt{x})^2dx}{2\sqrt{x}}$

Ex. 17.—*Ans.*

$$-\frac{2x(a^4+a^2x^2+x^4)dx+(a^2-x^2)(2a^2x+4x^3)dx}{(a^4+a^2x^2+x^4)^2}$$

Ex. 18.—*Ans.* $\dfrac{nx^{n-1}dx}{(1+x^n)^2}$

Ex. 19.—*Ans.* $\dfrac{4xdx}{(1-x^2)^2}$

Ex. 20.—*Ans.* $\dfrac{2dx}{\sqrt{1+-x^2}(\sqrt{1+x}+\sqrt{1-x})^2}$

Ex. 21.—*Ans.* $nm(a+bx^n)^{m-1}bnx^{n-1}dx.$

Ex. 22.—*Ans.* $\dfrac{3ax^2dx}{2\sqrt{ax^3}},$

Ex. 23.—*Ans.* $\frac{1}{2}x^{-\frac{1}{2}}y^{-\frac{1}{2}}dx-\frac{1}{2}x^{\frac{1}{2}}y^{-\frac{3}{2}}dy.$

Ex. 24.—*Ans.* $-mx^{-2}dx+\frac{3}{2}(x^2y)^{\frac{1}{2}}(2xydx+x^2dy).$

Ex. 25.—*Ans.* $2xy^{\frac{1}{2}}dx+\frac{1}{2}x^2y^{-\frac{1}{2}}dy-dz.$

Ex. 26.—*Ans.* $bcdx-\frac{3}{4}y^{-\frac{3}{4}}dy.$

Ex. 27.—*Ans.* $-3x^{-4}dx-y^{-2}dy+3z^2dz.$

Ex. 28.—*Ans.* $\frac{2}{5}(ax-y)^{-\frac{3}{5}}(adx-dy).$

Ex. 29.—*Ans.* $\frac{1}{2}(b-c)(x-y)^{-\frac{1}{2}}(dx-dy).$

Ex. 30.—*Ans.* $\dfrac{(4x^{\frac{3}{2}}+a)dx}{4\sqrt{x}\sqrt{x^2+a\sqrt{x}}}$

ARTICLE 24.

Ex. 5.—*Ans.* $\frac{1}{a^2}-\frac{2}{a^3}x+\frac{3}{a^4}x^2-\frac{4}{a^5}x^3+$, etc.

Ex. 6.—*Ans.* $a^{\frac{2}{3}}+\frac{2}{3}a^{-\frac{1}{3}}x-\frac{1}{9}a^{-\frac{4}{3}}x^2+\frac{8}{162}a^{-\frac{7}{3}}x^3-$, etc.

Ex. 7.—*Ans.* $a+\frac{1}{2}ba^{-1}x-\frac{1}{8}b^2a^{-3}x^2+\frac{3}{48}b^3a^{-5}x^3-$, etc.

Ex. 8.—*Ans.* $\frac{1}{c}+\frac{1}{2}c^{-3}y+\frac{3}{8}c^{-5}y^2+\frac{15}{48}c^{-7}y^3+$, etc.

Ex. 9.—*Ans.* $a^{-\frac{4}{3}}+\frac{2}{3}a^{-\frac{10}{3}}x^2+\frac{5}{9}a^{-\frac{16}{3}}x^4+\frac{40}{81}a^{-\frac{22}{3}}x^6+$, etc.

ARTICLE 26.

Ex. 4.—*Ans.* $x^n-nx^{n-1}y+\frac{n(n-1)}{2}x^{n-2}y^2$

$-\frac{n(n-1)(n-2)}{2\cdot3}x^{n-3}y^3+$ etc.

Ex. 5.—*Ans.* $x^{\frac{3}{2}}-\frac{2}{3}x^{\frac{1}{2}}y+\frac{3}{8}x^{-\frac{1}{2}}y^2+\frac{3}{48}x^{-\frac{3}{2}}y^3-$etc.

Ex 6.—*Ans.* $x^{-1}+x^{-2}y+x^{-3}y^2+x^{-4}y^3+$etc.

Ex. 7.—*Ans.* $a(x^{\frac{2}{3}}+\frac{2}{3}x^{-\frac{5}{3}}y+\frac{10}{18}x^{-\frac{8}{3}}y^2+\frac{80}{162}x^{-\frac{11}{3}}y^3+$etc.)

ARTICLE 32.

Ex. 8.—*Ans.* A maximum when $x=-7$; a minimum when $x=-5$

Ex. 9—*Ans.* A minimum when $x=\frac{b}{2}$.

Ex. 10.—*Ans.* A maximum when $x=\frac{b^3}{2c}$.

Ex. 11.—*Ans.* A minimum when $x=\pm\frac{b^2}{3a}$.

Ex. 12.—*Ans.* A maximum when $x=\frac{2b}{3c}$.

ARTICLE 33.

Ex. 13.—*Ans.* The height of the cylinder is equal to the diameter of the base.

Ex. 14.—*Ans.* The base of the triangle will be four-thirds of the abscissa of the extreme point.

ADDITIONAL EXAMPLES FOR PRACTICE.

DIFFERENTIALS OF FUNCTIONS.

Ex. 1. What is the differential of the function

$$u=x(a+x)(a^2+x^2)?$$

Ans. $du=(a^3+2a^2x+3ax^2+4x^3)dx$.

Ex. 2. What is the differential of the functio

$$u=(a+bx)^2(m+nx)^3\ ?$$

Ans. $du=3n(a+bx)^2(m+nx)^2dx+2b(a+bx)(m+nx)^3dx$

Ex. 3. Differentiate $u=(a+bx^2)^3(c+ex^4)^5$

Ans.

$du=20(a+bx^2)^3(c+ex^4)^4ex^3dx+6b(c+ex^4)^5(a+bx^2)^2xdx$

Ex. 4. Differentiate $u=(a+\sqrt{x})^3$

Ans. $du=\frac{3(a+\sqrt{x})^2dx}{2\sqrt{x}}$

Ex. 5. Differentiate $u=x(a^2+x^2)\sqrt{a^2-x^2}$

Ans. $du=\frac{(a^4+a^2x^2-4x^4)dx}{\sqrt{a^2-x^2}}$

Ex. 6. Differentiate $u=\frac{a+x}{\sqrt{a-x}}$

Ans. $du=\frac{3a-x}{2(a-x)^{\frac{3}{2}}}dx$

Ex. 7. Differentiate $u=\frac{x}{x+\sqrt{1-x^2}}$

Ans. $du=\frac{dx}{\sqrt{1-x^2}+2x(1-x^2)}$

Ex. 8. Differentiate $u=a+\frac{4\sqrt{x}}{3+x^2}$

Ans. $du=\frac{6(1-x^2)dx}{(3+x^2)^2\sqrt{x}}$

Ex. 9. Differentiate $u=\frac{x^n}{(1+x)^n}$

Ans. $du=\frac{nx^{n-1}dx}{(1+x)^{n+1}}$

Ex. 10. Differentiate $u=(a-x)\sqrt{a^2+x^2}$

Ans. $du=-\frac{(a^2-ax+2x^2)dx}{\sqrt{a^2+x^2}}$

Ex. 11. Differentiate $u=(a^2-x^2)\sqrt{a+x}$

Ans. $du=\frac{a^2-4ax-5x^2}{2\sqrt{a+x}}$

Ex. 12. Differentiate $u=\frac{ay}{\sqrt{x^2+y^2}}$

Ans. $du=-\frac{axydx-ax^2dy}{(x^2+y^2)^{\frac{3}{2}}}$

Ex. 13. Differentiate $u=(2a^2+3x^2)(a^2-x^2)^{\frac{3}{2}}$

Ans. $du=-15x^3\sqrt{a^2-x^2}.dx$

Ex. 14. Differentiate $u=\dfrac{\sqrt{a+x}+\sqrt{a-x}}{\sqrt{a+x}-\sqrt{a-x}}$

Ans. $du=-\dfrac{adx}{a\sqrt{a^2-x^2}-a^2+x^2}$

Ex. 15. Differentiate $u=\dfrac{a+2bx}{(a+bx)^2}$

Ans. $du=\dfrac{-2b^2xdx}{(a+bx)^3}$

DEVELOPMENT INTO SERIES.

Ex. 16. Develop into a series the function

$$u=\sqrt{a^2+x^2}$$

ns. $u=a+\dfrac{x^2}{2a}-\dfrac{x^4}{2\,.\,4a^3}+\dfrac{3x^6}{2\,.\,4\,.\,6a^5}-$etc.

Ex. 17. Develop into a series the function

$$u=\sqrt{2x-1}$$

Ans. $u=\sqrt{-1}(1+x-\dfrac{x^2}{2}+\dfrac{x^3}{2}-$etc.)

Ex. 18. Develop into a series the function

$$u=\frac{1}{\sqrt{b^2-x^2}}$$

Ans. $u=b^{-1}+\frac{1}{2}b^{-3}x^2+\dfrac{1\,.\,3}{2\,.\,4}b^{-5}x^4+\dfrac{1\,.\,3\,.\,5}{2\,.\,4\,.\,6}b^{-7}x^6+$etc.

Ex. 19. Develop into a series the function

$$u=(a^2+x^2)^{\frac{5}{3}}$$

Ans. $u=a^{\frac{10}{3}}+\frac{5}{3}a^{\frac{4}{3}}x^2+\dfrac{5\,.\,2}{3\,.\,6}a^{-\frac{2}{3}}x^4-\dfrac{5\,.\,2\,.\,1}{3\,.\,6\,.\,9}a^{-\frac{8}{3}}x^6+$etc.

Ex. 20. Develop into a series the function

$$u=\frac{1}{\sqrt[4]{a^4+x^4}}$$

Ans.

$$u=\frac{1}{a}-\frac{x^4}{4a^5}+\frac{5x^8}{4\cdot 8a^9}-\frac{5\cdot 9x^{12}}{4\cdot 8\cdot 12a^{13}}+\frac{5\cdot 9\cdot 13x^{16}}{4\cdot 8\cdot 12\cdot 16a^{17}}-\text{etc.}$$

MAXIMA AND MINIMA.

Ex. 21. Find the values of x which will render u a maximum or a minimum in the equation

$$u=x^4-8x^3+22x^2-24x+12$$

Ans. A maximum when $x=2$; a minimum when $x=1$ or 3.

Ex. 22. Find the values of x which will render u a maximum or minimum in the equation

$$u=3x^4-16x^3+6x^2+72x-1$$

Ans. There is a maximum when $x=2$ and a minimum when $x=-1$ or $+3$.

Ex. 23. Find the maximum or minimum values of

$$u=\frac{a^2x}{(a-x)^2}$$

Ans. There is a minimum value of u when $x=-a$.

Ex. 24. Find the maximum or minimum value of

$$u=4x^2-3x$$

Ans. There is a maximum when $x=\frac{3}{8}$.

Ex. 25. Find the fraction that exceeds its cube by the maximum quantity.

Ans. $+\frac{1}{\sqrt{3}}$

Ex. 26. What is the greatest positive value of $x-x^2$?

Ans. $\frac{1}{2}$.

Ex. 27. The hypothenuse and one side of a triangle are together equal to 18 inches. What is the length of each when the triangle is a maximum?

Ans. Hyp. 12 in.; side, 6 in.

Ex. 28. Required the minimum triangle formed by the radii produced and the tangent to the quadrant of a circle?

Ans. The triangle is isosceles.

Ex. 29. What value of x will render $57\sqrt{x}-x$ a maximum?

Ans. $812\frac{1}{4}$.

Ex. 30. A vessel is on the equator, 100 miles west of a certain meridian, sailing east at the rate of 7 miles per hour; another vessel is on that meridian, 50 miles north of the equator, at the same moment sailing south at the rate of 10 miles per hour; where will the vessels be when nearest to each other?

Ans. The first will be 43.63 miles, nearly, west of the meridian, and the other 30.5 miles, nearly, south of the equator; and the distance between them will be 53.14 miles, nearly.

Ex. 31. Three towns are at the angles of a right-angled triangle; A and B at the base, which is 25 miles long from east to west, and C 100 miles north of A. There is a railway from A to C, on which the cars go through in 5 hours. A man at B wishes to reach C in the shortest time, and can travel 5 miles per hour on horseback. He decides to ride to the railway and take the cars. At what point must he meet the train in order to accomplish his object?

Ans. At a point $\frac{25}{\sqrt{15}}$ miles from A.

Ex. 32. Let there be two lines, each equal to b, standing at the extremities of a third line a, and equally inclined to it; what must be the length of the fourth line C, connecting the upper extremities of the lines b, so that the area of the trapezium thus formed shall be a maximum?

Ans. $C=\frac{a}{2}+\sqrt{2b^2+\frac{a^2}{4}}$

Ex. 33. The two sides, each equal to b, of an isosceles triangle being given, what must be the length of the third side, so that the area of the triangle shall be a maximum?

Ans. $\sqrt{2b^2}$

Ex. 34. What decimal fraction exceeds its cube more than any other numerical *quantity* can exceed *its* cube?

Ans. .577+

Ex. 35. Divide 25 into two such parts that the product of the second power of one part by the third power of the other may be a maximum; and which part must be cubed?

Ans. The numbers are 15 and 10.
The number 15 must be cubed.

Ex. 36. What will be the proportions of the largest rectangle that can be inscribed in an ellipse of which $2A$ is the transverse axis and $2B$ the conjugate?

Ans. The sides of the rectangle are $A\sqrt{2}$ and $B\sqrt{2}$

Ex. 37. What is the height of the smallest cone that will enclose a sphere, lying on the plane of its base?

Ans. Four times the radius of the sphere.

Ex. 38. What are the greatest and least values of the ordinate of the curve represented by

$$a^2y=ax^2-x^3$$

Ans. When $x=\frac{2}{3}a$, y is a maximum; and when $x=o$, y is a minimum, being also equal to zero.

SUBTANGENTS AND SUBNORMALS.

Ex. 39. What is the subnormal to the curve whose equation is

$$y^2 = 2a^2 \log. x$$

Ans. $\dfrac{a^2}{x}$

Ex. 40. What is the value of the subnormal of the curve whose equation is

$$2ay^2 + a^3 = 2x^3$$

Ans. $\dfrac{3x^2}{2a}$

Ex. 41. What is the value of the subtangent of the curve whose equation is

$$y^2 = \frac{x^3}{a-x}$$

Ans. $\dfrac{2x(a-x)}{3a-2x}$

Ex. 42. Required the subtangent of the curve whose equation is

$$xy^2 = a^2(a-x)$$

Ans. $-\dfrac{2(ax-x^2)}{a}$

Ex. 43. When is the subtangent of the preceding curve a minimum? *Ans.* When $x = \frac{1}{2}a$

Ex. 44. Required the value of the subtangent of the curve whose equation is

$$x^2y^2 = (a+x)^2(b^2-x^2)$$

Ans. $-\dfrac{x(a+x)(b^2-x^2)}{x^3+ab^2}$

CURVATURE AND CURVED LINES.

Ex. 45. Find the radius of curvature at any point of the cubical parabola whose equation is

$$y^3 = ax$$

Ans. $R = \dfrac{(9y^4+a^2)^{\frac{3}{2}}}{6a^2y}$

Ex. 46. When is the curvature of the preceding curve greatest?

Ans. When $y=\sqrt[4]{\dfrac{a^2}{40.5}}$

Ex. 47. Find the radius of curvature at any point of the logarithmic curve whose equation is

$$y=a^x$$

Ans. $R=\dfrac{(M^2+y^2)^{\frac{3}{2}}}{My}$, M being the modulus and a the base.

Ex. 48. Find the point of greatest curvature of the logarithmic curve.

Ans. The point whose ordinate is equal to $\dfrac{M}{\sqrt{2}}$

Ex. 49. Find whether the curve whose equation is

$$y^3=x^5$$

has a point of inflexion.

Ans. The curve has a point of inflexion at the origin.

Ex. 50. Find the point of inflexion in the curve whose equation is $ax^2=a^2y+x^2y$

Ans. There is an inflexion at the point where $y=\frac{1}{10}a$.

Ex. 51. Find the point of inflexion in the curve whose equation is $x^2y^2=a^2(ax-x^2)$

Ans. There are two points of inflexion corresponding to $x=\dfrac{3a}{4}$, and $y=\pm\dfrac{a}{\sqrt{3}}$

Ex. 52 Find whether the curve whose equation is

$$(y-b)^3=(x-a)^2$$

has a cusp at the point where the tangent is parallel to the axis of ordinates.

Ans. There is a cusp of the first order at the point where $y=b$ and $x=a$

INTEGRAL CALCULUS.

INTEGRATION OF DIFFERENTIALS.

Ex. 1. Integrate the differential

$$du=\frac{dx}{(a-x)^5}$$

Ans. $u=\frac{1}{4(a-x)^4}$

Ex. 2. Integrate the differential

$$du=\frac{4xdx}{(1-x^2)^2}$$

Ans. $u=\frac{2}{1-x^2}$

Ex. 3. Integrate the differential

$$du=\frac{2adx}{x\sqrt{2ax-x^2}}$$

Ans. $u=-\frac{2\sqrt{2ax-x^2}}{x}$

Ex. 4. Integrate the differential

$$du=\frac{x^8dx}{\sqrt{a^9+6x^9}}$$

Ans. $u=\frac{\sqrt{a^9+6x^9}}{27}$

Ex. 5. Integrate the differential

$$du=\frac{dx}{\sqrt{1+x^2}}$$

Ans. $u=x-\frac{x^3}{2\cdot3}+\frac{3x^5}{2\cdot4\cdot5}-\frac{3\cdot5x^7}{2\cdot4\cdot6\cdot7}+\text{etc.}$

Ex. 6. Integrate the differential

$$du=\frac{3x^2dx}{x^3+a^2}$$

Ans. $u=log.\,(x^3+a^2)$

Ex. 7. Integrate the differential

$$du=\frac{5x^3dx}{3x^4+7}$$

Ans. $u=\frac{5}{12}\,log.\,(3x^4+7)$

Ex. 8. Integrate the differential

$$du=\frac{3x^2+2x+1}{x^3+x^2+x+1}dx$$

Ans. $u=log.\,(x^3+x^2+x+1)$

Ex. 9. Integrate the differential

$$du=\frac{dx}{x^2\sqrt{a^2+x^2}}$$

Ans. $u=-\frac{\sqrt{a^2+x^2}}{a^2x}$

SOME EXAMPLES

SHOWING THE APPLICATION

OF THE

PRINCIPLE SET FORTH IN THIS WORK.

The grand instrument by which the ancient mathematicians brought the science of Geometry to so great a degree of perfection was the "*method of exhaustion*," or otherwise, the "*reductio ad absurdum*." When the science awoke from its long sleep, by the brilliant discoveries of Cavalieri, Pascal, Descartes, Leibnitz and Newton, the mathematical world became impatient of the scrupulously exact but tedious and laborious method of the ancients, and demanded shorter and easier methods. Various were the methods resorted to. The indivisibles of Cavalieri was one of them, being pronounced by Carnot to be "an abbreviation by means of which we obtain promptly and easily, in many cases, what could be discovered only by long and painful processes according to the method of exhaustion." Such, also, were the infinitesimals of Pascal and Leibnitz, by which they converted curves into polygons, and thus avoided the "reductio ad absurdum," by the adoption of a notion, of which the absurdity is intrinsic and without any alternative. Such also was the object of the celebrated lemmas of Newton, in the first book of his Principia; of which he says: "These lemmas are premised to avoid the tediousness of deducing perplexed demonstrations *ad absurdum* according to the method of the ancient geometers." The first of these lemmas (see page 40) seems to be a sort of generalization of the *reductio ad absurdum* so that the principle may be applied at pleasure without going through the tedious formula in each particular case.

Many modern geometers have adopted the infinitesimal method, and demonstrate the properties of the circle, for instance, by considering it a polygon with an indefinite number of sides. Says one * "the circumference is the limit of the perimeter of the polygon," and adds, "no sensible error can arise in supposing that what is true of such a polygon is also true of its limit, the circle." But can *any* error arise? If so, then the polygon does not, strictly speaking, coincide with the circle.

Another author has struck out a different method. He says, indeed truly, that, † "strictly speaking, the circle is not a polygon, and the circumference is not a broken line," and therefore instead of the method of infinitesimals he adopts what he terms the method of limits. ‡ "The principle," says he, "upon which all reasoning by the method of limits is governed, is that, *whatever is true up to the limit is true at the limit.* We admit this as an axiom of reasoning because we cannot conceive it to be otherwise." Again, ‖ "Now, it is evident that by the process described" (bi-secting the sides indefinitely) "the polygons can be made to approach as nearly as we please to equality with the circle, but can never entirely reach it." Bearing in mind this last admission, how shall we construe the above "axiom"? How can we follow the polygon up *to* the limit without arriving *at* the limit? But if "up to the limit" means anything short of "at the limit," then truly the polygon does not become the circle, and we

* Davies' Legendre, revised edition of 1856.

† Ray's Plane and Solid Geometry, Art. 477. ‡ Art. 201. ‖ Art. 475.

must deny that "whatever is true" (of the polygon) "up to the limit, is true" (of the circle) "at the limit." We say "of the circle" because the circle *is* the limit of the polygon. In this case all the original differences between the polygon and the circle remain intact. The boundary of a polygon is a broken line, that of a circle is not. The periphery of a polygon has angles, that of a circle has none. The area of the polygon is less than that of the circumscribed circle, which is not true of the circle itself.

"The author does not claim the credit of having discovered or invented this new axiom. "In explaining the doctrine of limits," says he, 'the axiom by Dr. Whewell is given in the words of that eminent scholar.' Now, Dr. Whewell certainly had no use whatever for any such axiom. For, according to his view, the variable magnitude not only approaches as near as we please, but actually reaches its limit. Thus, says he, 'a line or figure *ultimately coincides* with the line or figure which is its limit.' Now, most assuredly, if the inscribed polygon ultimately coincides with the circle, then no new axiom is necessary to convince us that what is always true of the polygon is also true of the circle. For this is only to say that what is true of the variable polygon in all its forms is true of it in its last form. According to his view, indeed, there was no chasm to be bridged over or spanned, and consequently there was no need of any very great labor to bridge it over or to span it. The truth is, however, that although he said the two figures would ultimately "coincide," leaving no chasm between them to be crossed, he felt that there would be one, and hence the new axiom for the purpose of bridging it over.

But if we reject these methods of getting rid of the "reductio ad absurdum," what shall we do? "If we deny that the polygon and the circle ever coincide, how shall we

* Bledsoe's Philosophy of Mathematics.

bridge over the chasm between them so as to pass from the knowledge of right lined figures and volumes to that of curves and curved surfaces? Shall we, in order to bridge over this chasm, fall back on the *reductio ad absurdum* of the ancients? or can we find a more short and easy passage without the sacrifice of a perfectly logical rigor in the transit? This is the question. This is the very first problem which is and always has been presented to the cultivators of the modern methods. Is there, after the lapse and labor of so many ages, no satisfactory solution of this primary problem? It is certain that none has yet been found which has become general among mathematicians." I believe, moreover, that such a method never will be found. But I also believe that no "bridge" is necessary—that instead of starting with the polygon and, with it, trying to reach the circle, we may go directly to the circle itself and boldly compel it to yield up its properties to the open light of mathematical truth without any "sacrifice of perfectly logical rigor."

THE

DIRECT METHOD OF RATES

AS APPLIED TO

PLANE AND SOLID GEOMETRY.

Several of the propositions in Elementary Geometry have never been solved directly. The result has been obtained, either by the "*reductio ad absurdum*," which is logical and conclusive, although indirect; or else by the method of limits, or infinitesimals.

I propose to apply the *direct method of Rates* to some of these propositions, by way of example and illustration. For this purpose we assume the following fundamental principles or axioms.

Axiom I.

The rate of increase of any geometrical magnitude while being generated, may be measured at any moment, by a suppositive uniform increment that would arise if the generatrix were to continue to move with the existing velocity, measured in a direction perpendicular to itself, for one unit of time without changing its own magnitude.

Thus a right cone may be generated by its base flowing upward and gradually contracting so as to bring it to a point at the vertex of the cone. But the *rate* at which it might be increasing at any instant would be represented by the cylinder that would be formed if the generatrix were to continue its movement, without contraction, for one unit of time; for the base of this cylinder would represent the generatrix and its height would represent its velocity—and these two, together, compose the rate.

Axiom II.

If two magnitudes begin to be at the same moment, and the ratio of their rates of increase is constant, the ratio of the magnitudes, themselves, will be constantly the same as that of their rates.

Thus, if two persons set out at the same moment and place to travel in the same direction, at constant rates, the ratio of the distances traveled by each will be constantly the same as that of their rates of travel.

Axiom III.

If a geometrical magnitude be generated at a uniform rate, the quantity generated will be proportional to the time occupied.

This axiom needs no illustration.

Proposition I.

Two angles are to each other as the intercepted arcs, described from their vertices with equal radii.

Let A C B and D C B be the two given angles, and A B and D B the corresponding arcs. Since the arcs are described with equal radii they may be considered as parts of the same circumference, and the angles as being at the center of the same circle, as in the figure. Let us suppose the radius C A to revolve around the center C at a uniform rate, thus describing the angles and the arcs. Then it is evident (Axiom III) that the arcs will be to each other as the times in

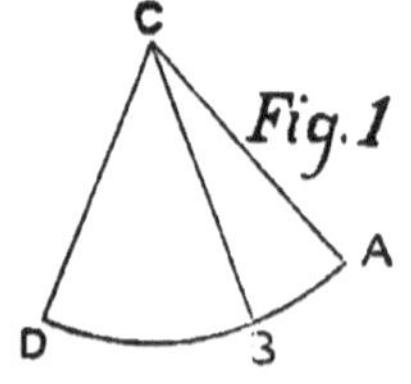

which they are described, and the angles, being described during the same times will be in the same ratio. Hence, the angles will be to each other as the corresponding arcs.

Proposition II.

Rectangles having the same altitude are to each other as their bases.

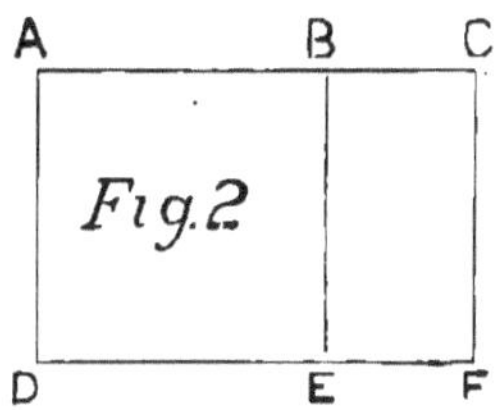

Let A E and B F be the two rectangles, whose bases may be placed on the same line, as in the figure. Then B E will be the height common to both. Suppose the rectangle A E to be generated by the line B E moving towards A D at a uniform rate and constantly parallel to itself; and the rectangle B F to be generated by the same line B E moving in like manner toward C F at precisely the same uniform rate. Then (Axiom III) the areas A E and B F will be to each other as the times required to described them; and the bases of the rectangles, which measure the distances passed over by the generatrix in each, being described at a uniform rate, will be proportional to the same times. Hence, the areas of the two rectangles are to each other as their bases.

Proposition III.

Two rectangular parallelopipeds having equal bases are to each other as their altitudes.

Suppose the volumes to be described or generated by their bases flowing upward at equal, uniform rates; then (Axiom III) the volume generated in each case will be proportional to the time occupied by the movement of its generatrix; and the altitude, which measures the movement in each case will also be proportional to the time. Hence the volumes will be to each other as their altitudes.

Proposition IV.

The volumes of two Pyramids having equivalent bases and equal altitudes are equivalent to each other.

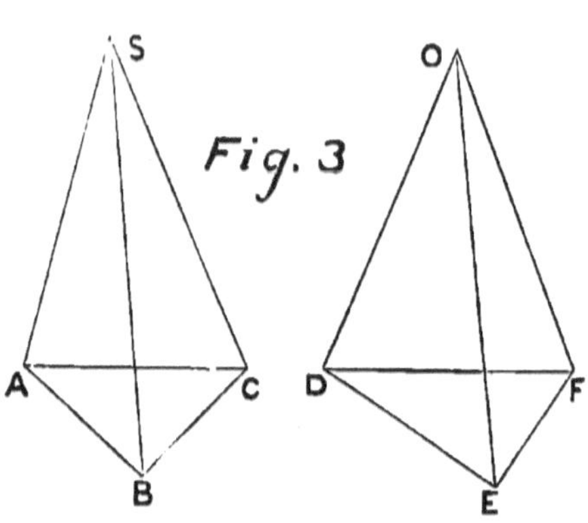

Fig. 3

Let S A B C and O D E F be two pyramids of which the bases A B C and D E F are equivalent triangles and the altitudes are equal. Then let us suppose each of these pyramids to be generated by the flowing of its base upward at the same rate as the other, in a vertical direction, and contracting its sides so that its angles shall always be in the edges of the pyramid. Then, since the rate of generation of the pyramid at any moment will be in proportion to the area of the generatrix, multiplied by the rate at which it is moving; and since, in this case, the rates are the same and the generatrices at all times equivalent, the rates of generation of the two pyramids will, at all times, be equal; hence (Axiom II), the magnitudes of the parts generated will be constantly equivalent, and, of course, the pyramids, when completed, will also be equivalent.

Proposition V.

To find the area of a Circle.

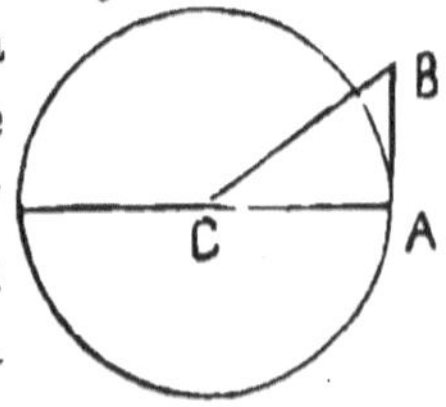

We will suppose the circle to be generated by the revolution of radius C A about the center C at a uniform rate. When the radius is in the position C A, and revolving toward B, every point in it will *tend* to move in a direction perpendicular to C A, and hence the point A will *tend* to describe the line A B tangent to the circle and perpendicular to C A; and if left to its ten-

Fig. 4.

dency *would* describe that line at a uniform rate. The line, therefore, may be taken as the *symbol* representing the rate of increase or generation of the circumference of the circle.

But while the point A tends to move in the direction A B, every point in the radius C A *tends* to move in a direction parallel to it, and at a rate proportional to its distance from the center C. Hence the radius itself, *if left to its tendency* when at C A, would be found at C B, when the point A is at B ; and the triangle C A B would be generated at a uniform rate during the same time that the line A B is generated. The triangle may, therefore, be taken as the symbol of the rate at which the area of the circle is generated, and the ratio of these symbols is also the ratio of the rates which they represent. But the triangle is equal to $\frac{1}{2}$ C A . A B; that is, the ratio between the rate of generation of the circumference and that of the area of the circle is half radius ; and this, being constant, is the ratio between any part of the circumference and the corresponding part of the circle throughout their generation, and, of course, of the entire circumference to the entire circle. Hence the area of the circle is equal to half the radius into the circumference.

Proposition VI.

To find the area of the convex surface of a Cone.

Suppose the cone to be generated by the revolution of the triangle A D C (Fig. 5) about the axis D C. The hypothenuse A D will generate the convex surface, and the point A will generate the circumference of the base. When the triangle is in the position A D C and revolving towards E, the point A if left to its *tendency* at that instant would describe the line A E, perpendicular to A C, in some unit of time, and hence

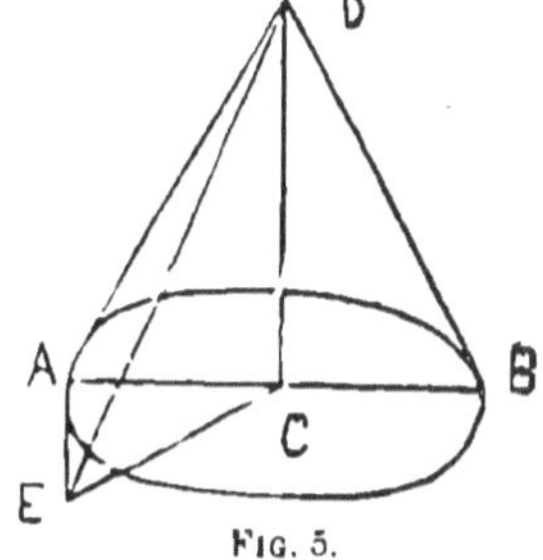

Fig. 5.

A E may be taken to represent the *rate* at which the circumference of the base is generated. Now every point in the line A D tends to move in a direction parallel to A E, and at a rate proportional to its distance from the axis D C; hence if left to that tendency the line A D would describe the triangle A D E, and be found at D E in the same unit of time. Hence A D E (Axiom I) may be taken to represent the corresponding rate of generation of the convex surface of the cone. But $A D E = A E . \frac{AD}{2}$ or $\frac{A D E}{A E} = \frac{A D}{2}$, that is, the ratio between the rates of generation of the convex surface of the cone and the circumference of its base is constant and equal to half its slant height. Hence the ratio between the magnitudes generated will be the same (Axiom II), and the convex surface divided by the circumference of the base equals half the slant height, or, the convex surface equals the circumference of the base multiplied by half the slant height.

Proposition VII.

To find the measure of the volume of a Cone.

The cone being supposed to be generated by the revolution of a right angled triangle about one of its sides, which becomes the axis of the cone, while the base is generated by the other side as its radius, let us suppose the generating triangle to have arrived at the position A D C (Fig. 5)—the point A moving towards E. Then every point in the triangle *tends* to move in a direction perpendicular to its plane and at a rate proportional to its distance from the axis C D; so that if A E is taken to represent the line that would be described by the point A in a unit of time in consequence of that *tendency*, then at the end of the same unit of time the line A D would be found at E D, and the triangle A D C, at E D C, so that the pyramid D A E C would be the volume generated by the triangle, during the same unit of time and may therefore (Axiom I) be taken to represent the *rate* at which the cone is generated; while, at the same time, the triangle A C E would be described by the radius A C of the base, and would, therefore, represent the rate at which the base of the cone was generated. But the volume of the pyramid D A E C is equal to its base A C E multiplied by one-third of its altitude D C. Hence the rate of generation of the cone divided by that of its base = a constant quantity. Therefore (Axiom II) the cone itself divided by its base is equal to the same quantity, being one-third of its height—or the volume of the cone is equal to its base multiplied by one-third of its height.

Lemma I.

To find the area of the convex surface of an ungula with a semicircular base.

An ungula of this kind is formed by passing a plane obliquely through the center of the circular base of a right cylinder. The ungula, thus cut off, will have, for its lower base, half of the base of the cylinder, and, for the upper one, a semi-ellipse, while the elements of its convex surface will be portions of the elements of the cylinder. The height of the ungula will be measured by that element of its convex surface which is ninety degrees from the extremities of its base.

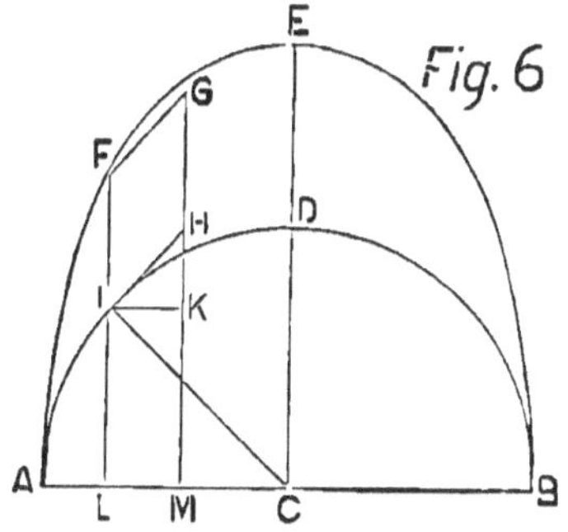

Let E D A B be the convex surface of the ungula of which A D B is the base and E D, perpendicular to the base, is the height. Then suppose the convex surface to be generated by an element F I, moving parallel to itself, with its foot in the circumference of the circular base, and its length varying in such a manner that its upper extremity shall be constantly in the elliptic boundary of the upper base. Draw I H tangent to the base at the foot of the element F I, F G parallel to I H, G H parallel to F I, I L and H M perpendicular to the diameter A B, and I K parallel to it. Join I C and assume I H to be the line that *would be* generated by the foot of the element, F I, in one unit of time, if left to its tendency, with the velocity it had the moment it arrived at its present position; it will then (Axiom I) represent the rate at which the circumference of the base is generated at that moment; and the rectangle F G H I, being generated at a uniform rate in the same unit of time, will be the symbol of the rate at which the convex surface of the ungula is generated. Also L M, being a uniform increment of the diameter of the base, generated in the same unit of time, will represent the rate at which that diameter is generated.

Since every plane cutting A B at right angles, will cut the surfaces of the ungula in similar triangles, there will be a constant ratio, which we call r, between each element of the convex surface of the ungula and the corresponding ordinate of the base; hence we may make $IL = r \,.\, FI$. Since the triangles I H K and I L C are similar, we have

$$IL : IK :: IC : IH$$

whence $$IL \,.\, IH = IK \,.\, IC \text{ or } r \,.\, FI \,.\, IH = IK \,.\, R$$

substituting for I K, its equal L M and dividing by r we have

$$FI \,.\, IH = \frac{R}{r} LM = ED \,.\, LM.$$

The first member of this equation is equal to the rectangle F G H I, which represents the rate at which the convex surface is generated, and the last member is composed of the constant E D into L M which represents the rate at which the diameter is generated. Hence, the rates of generation of the convex surface of the ungula, and of the diameter of its base, having a constant ratio from the beginning, their magnitudes, when completed will (Axiom II) have the same ratio; therefore the convex surface of the ungula will be equal to the *diameter of the base multiplied by its extreme height.*

Lemma II.

To find the measure of the volume of an ungula with a semicircular base.

Let E A D B be the ungula, of which the semicircle A D B is the base and E D, perpendicular to D C is the extreme altitude. Let F I represent one of the elements of the convex surface, and join F and I to C. Then we may suppose the ungula to be generated by the vertical right-angled triangle C F I revolving about the center C, its base being one of the radii of the semicircle. and the side F I being always vertical, and forming one of the elements of the convex surface of the ungula. Draw I H to represent the rate of generation of the circumference of the base of the ungula ; construct the rectangle F G H I and join G and H to C, then (Axiom I) the triangle I H C will represent the rate of generation of the base, the rectangle F G H I will represent the rate of generation of the convex surface, and the pyramid C F G H I will represent the rate at which the volume of the ungula itself will be generated ; for while the point I would move to H under its *tendency* at the moment, the line F I would move to G H describing the rectangle, and the triangle C F I, would move to C G H describing the pyramid C F G H I, and these being all described simultaneously, at a uniform rate, will represent the relative rates of generation of the respective magnitudes to which they belong.

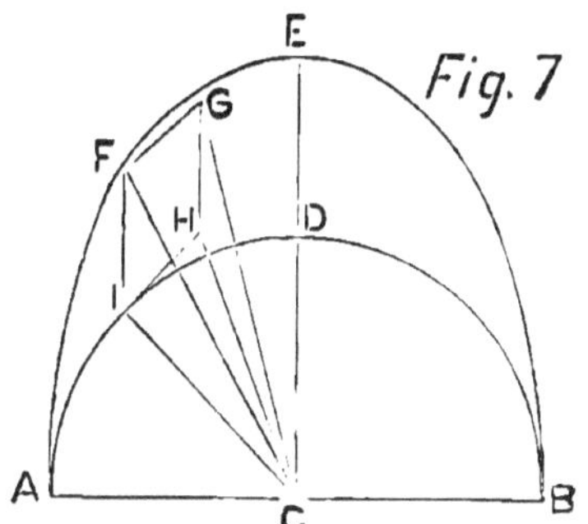

Now the volume of the pyramid is equal to its base F G H I multiplied by one-third of its altitude I C or R; or, in other words, the rate of generation of the volume of the ungula is equal to that of the convex surface multiplied by one-third of the radius of its base, which is therefore the constant ratio between these two rates, and hence also, (Axiom II) the ratio between the completed magnitudes. So that the volume of the ungula is equal to its convex surface multiplied by one-third of its radius of its base—that is (Lemma I), $ED \cdot AB \cdot \frac{1}{3}R = ED \cdot \frac{2}{3}R^2$—or, its extreme altitude multiplied by two-thirds of the square of the radius of its base.

Proposition VIII.

To find the area of the surface of a Sphere.

Suppose the sphere to be generated by the revolution of the semicircle C B D (Fig. 8) about the diameter C D. Then the semi-circumference C B D will generate the surface of the sphere. Now every point in the curve C B D tends to move in a direction perpendicular to its plane at a rate proportional to its distance from C D, the axis of revolution, and under this *tendency* it would in

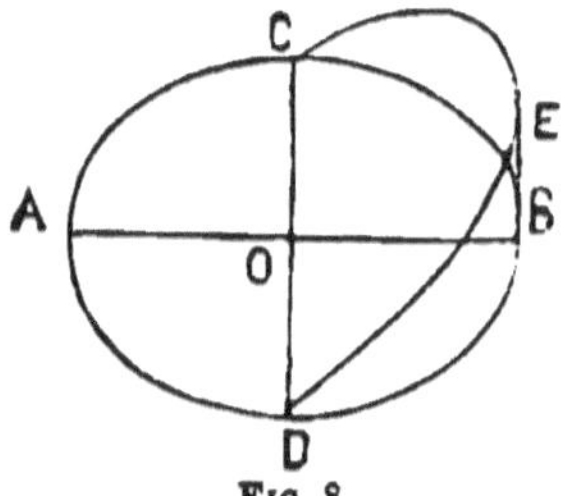

Fig. 8.

in some unit of time assume the position of the semi-ellipse C E D, generating at the same time the convex surface of the ungula C E D B, which is, therefore, the symbol (Axiom I) representing the rate of generation of the surface of the sphere, while the line E B described by the point B, in the middle of the arc C B D, and perpendicular to its plane is the symbol representing the corresponding rate of generation of the circumference of a great circle. But the convex surface of the ungula is equal to its extreme height E B, multiplied by the diameter C D, which is, therefore, the constant ratio between the rates of generation of the surface of the sphere and of the circumference of its great circle. The magnitudes themselves are, therefore (Axiom II), in the same ratio, and the surface of a sphere is equal to its diameter multiplied by the circumference of its great circle.

Proposition IX.

To find the measure of the volume of a Sphere.

The sphere being supposed to be generated by the revolution of the semicircle C B D about the diameter C D (Fig. 8), when it is revolving towards the point E, every point in it will *tend* to move in a direction perpendicular to its plane, and at a rate proportional to its distance from C D the axis of revolution; and the point B in the middle of the arc C B D will tend to describe the line B E perpendicular to the plane of C B D, in some unit of time, and would do so if left to that tendency. The semicircle C B D, at the end of the same unit of time, would be found in the ellipse C E D, having described or generated the ungula E C B D, which may, therefore (Axiom I), be taken as the symbol of the rate at which the volume of the sphere is generated, while E B is the symbol of the rate at which the circumference of its great circle is generated. But the volume of the ungula is equal to its extreme height E B multiplied by two-thirds of the square of the radius, which is, therefore, the ratio between the rates of generation of the volume of the sphere and of the circumference of its great circle. Hence the magnitudes themselves are in the same ratio (Axiom II) and the volume of the sphere is equal to the circumference of its great circle multiplied by two-thirds of the square of its radius—or what is the same thing, to its surface multiplied by one-third of its radius.

PROPOSITION X.

To find the area of an Ellipse.

Let there be an ellipse (Fig. 9) described on A B as its transverse axis and a circle described on the same line as a diameter. Suppose the ellipse to be generated by the flowing of the double ordinate E C from B towards A, and the circle, by its double ordinate D H flowing in the same direction; both lines being constantly together and remaining perpendicular to the axis A B. Then the ratio of the rates at which the surfaces are generated will be at all times the same as that of their respective generatrices—that is, as D L : E L : : R.O : M O : hence (Axiom II) the surfaces generated will have the same ratio. But the surface of the circle is equal to $\pi\ \overline{RO}^2$ or π A O . R O ; hence that of the ellipse will be π A O . M O ; or, π into the rectangle of the semi-axes.

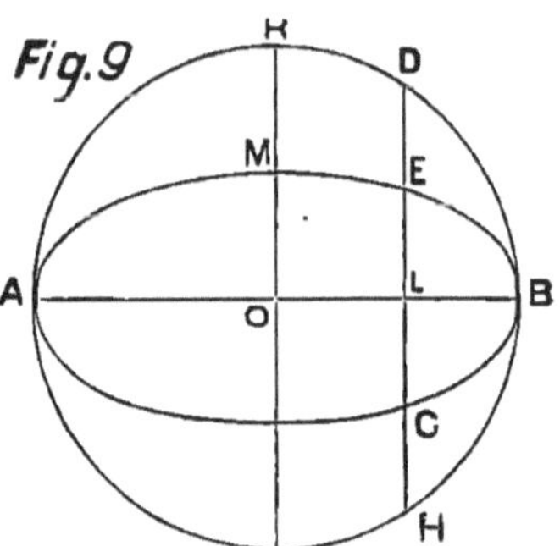

PROPOSITION XI.

To find the measure of the volume of an Ellipsoid.

Suppose the ellipsoid to be generated by the revolution of a semi-ellipse A M B (Fig. 9) about its major axis A B, and suppose a sphere A R B H to be described about the same axis as a diameter. We may then suppose each to be generated by a circle flowing with the same uniform velocity from B toward A,—the radii D L and E L of the circles varying so that the circumference of each shall pass through the meridian curve of the volume generated by it. Then, since the circles move with equal velocities, the ratio of the rates at which volumes are generated will be at all times the same as that of the generating circles, or as

$$\overline{DL}^2 : \overline{EL}^2 : : \overline{RO}^2 : \overline{MO}^2$$

and hence (Axiom II) the ratio of the volumes themselves will be the same.

But the volume of the sphere is

$$4/3 \, \pi \, \overline{RO}^3 \text{ or } 4/3 \, \pi \, AO . \overline{RO}^2$$

and therefore the volume of the ellipsoid is

$$4/3 \, \pi \, AO . \overline{MO}^2$$

or equal to $\frac{2}{3}$ of that of a cylinder whose altitude is the transverse axis and whose diameter is the conjugate axis of of the generating ellipse.

THE

DIRECT METHOD OF RATES

AS APPLIED TO

SOME PROBLEMS IN MECHANICS.

To find the Center of Gravity.

Definition.—The center of gravity of any body is that point, about which, if supported, the body will be evenly balanced, so that it will have no tendency, arising from its weight, to change its position.

It is proven in mechanics, that if the weights of several bodies be multiplied by the distances of their respective centers of gravity from a given vertical plane, the sum of the products will be equal to the product arising from multiplying the sum of the weights of the bodies by the distance of their common center of gravity from the same plane. Hence, if we divide the sum of these products by the sum of the weights, we obtain the distance of their common center of gravity from that plane.

The product of the weight of any body into the distance of its center of gravity from any fixed vertical plane is called the *moment* of that body with respect to that plane. We say therefore that the sum of the moments of all the bodies in any system is equal to the moment of their sum ; and also that the sum of the moments of the several parts of any single body is equal to the moment of the body itself.

This principle is useful in determining the centers of gravity of those bodies whose dimensions can be reduced to geometrical magnitudes. Among such bodies are some that can be divided into parts whose respective centers of gravity can be found by inspection, and thus the center of gravity of the whole is easily obtained; but there are many that cannot be thus divided, however small the parts may be. For this latter class it is necessary to resort to the ultimate particles of matter, so small that each one may be considered as occupying the same place as its center of gravity. It is, of course, impossible to estimate the weights of such particles singly, or the sum of the required products, by any direct process. Hence we resort to one that is indirect. That is, we find the *rate* at which the sum of these products increases, and from that rate determine the sum itself.

In the following examples, it is assumed that the lines and surfaces are homogenous, material bodies, of uniform thickness so small as not to affect the distance of the center of gravity from the plane of reference.

To find the center of gravity of a curved line.

Let B A C be the curved line symmetrical about the axis A G; then it is evident that its center of gravity will be somewhere in that axis, and that the center of gravity of each half of the curve will be at the same distance from the plane of reference, D E, as that of the whole.

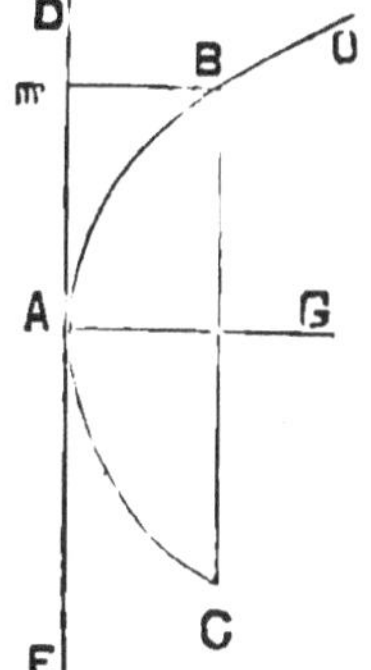

We will, therefore, take the part A B, and suppose it to be generated by a point flowing from A toward B around the curve at a rate determined by some law. From B draw a line B O tangent to the curve at B. Then the generating point on arriving at B would *tend* to flow at a rate, and in a direction, which, if uniformly continued, would generate a tangent line as B O in a unit of time; and, therefore, B O may be taken to represent the rate of increase of the curve at the point B:—that is, it will represent the differential of the curve. Calling the weight of the curve w, this would be represented by dw. Since the line B O represents a rate which is something which exists at the point B only, its moment will be expressed by B O × B m, or $x\ dw$, which represents therefore the moment of the rate of generation of the curve. Now, since B O or dw represents the rate of increase of the curve, it will, of course, represent that of the sum of its particles, and this rate multiplied by Bm or x ($x\ dw$) will represent, not the sum of the *products* of the particles into their respective distances from D E which we are seeking, but their rate of increase or the differential of that sum: so that the moment of the rate of increase of the curve is, also, the rate of increase of the sum of the moments of its particles, and its integral ($\int x\ dw$) will be the sum of these moments.

If, therefore, we divide $\int x\ dw$ by the mass of the curve, we shall have for a quotient the required distance of its center of gravity from the plane of reference.

NOTE.—It might be supposed that the moment of B O should be B O (x + ½ B O) since the center of gravity of B O is in the middle of its length. But it must be remembered that B O is not an integral part of the curve, but a *symbol*, or an *ideal* line representing a *supposition*. It is a measure of *value;* but it is a value existing at the point B, and there only, viz.: that of the rate of generatio i of the curve at the moment the generating point arrives at B. Similarly, if a falling body acquire, at a point *A*, a rate of motion of 50 feet in one second, we multiply its weight by 50 and obtain its momentum. Here the rate of motion, represented by 50 feet, and the corresponding momentum belong, wholly, to the point *A* where these values exist at the moment.

So the value of the line B O and its moment belong, wholly, to the point B where the rate exists whose value is represented by B O.

For another illustration: suppose a body to be generated in a direction from the plane of reference by successive accretions of one unit each; and suppose that the last unit had been generated at a rate that would have produced n such units in one unit of time, then n units will represent the rate of increase of the last unit, which is, also, that of the entire body; and the sum of the products of all the units into the respective distances of their centers of gravity from the plane of reference, will be increasing at the same rate as that of the last unit (represented by n such units) into its distance.

But this is nothing more than the moment of the rate of increase of the sum of the products, which is, therefore, equal to the rate of increase of the sum of the moments; and this divided by the sum of the parts, will give the rate of increase of the distance of the center of gravity of the whole from the plane of reference.

The same thing may be expressed more briefly as follows: As the body is generated and each successive increment is multiplied into its distance from the plane of reference, the rate at which the last product is formed will be equal to the rate at which the last increment is formed multiplied into its distance from the plane of reference, and will be the rate of increase of the sum of the products. Hence the moment of the rate of increase of the body is equal to the rate of increase of the sum of the moments.

To apply this formula we obtain the value of dw by taking the differential of the length of the line from its equation, and then multiply and integrate in the same way as in the infinitesimal system.

Proposition.

To find the center of gravity of a cycloid.

Let A C B be the cycloid of which C F is the axis. A B is the base and D E tangent to the curve at the upper extremity of the axis and parallel to the base. As the cycloid is symmetrical about its axis, the center of gravity will be in it, and the problem will be, to find its distance from C, as the origin.

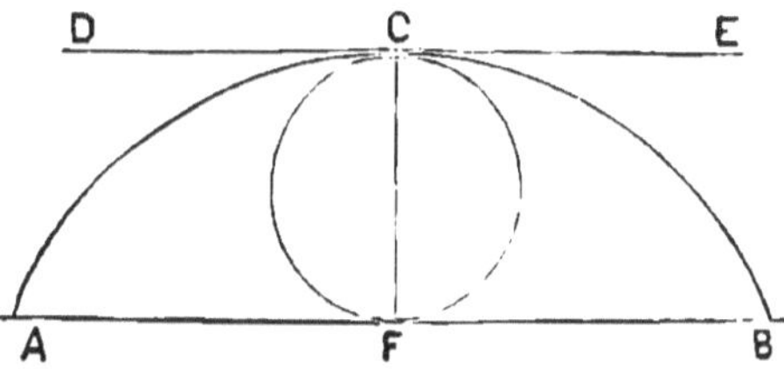

This distance will be expressed in terms of y, but as, in the equation of the curve, the value of y is reckoned from the line A B, we must, in order to reckon the distance from C, substitute for y in the differential equation of the curve, another y or y' which will be equal to $2r - y$; whence $y = 2r - y'$.

Making this substitution in the equation

$$dx = \frac{ydy}{\sqrt{2ry - y^2}} = dy\sqrt{\frac{y}{2r - y}}$$

we have

$$dx = d(2r - y')\sqrt{\frac{2r - y'}{y'}} = -dy'\sqrt{\frac{2r - y'}{y'}}$$

hence

$$dx^2 = dy'^2\frac{2r - y'}{y'}$$

and

$$dw = \sqrt{dx^2 + dy'^2} = dy'\sqrt{1 + \frac{2r - y'}{y'}} = \sqrt{\frac{2r}{y'}}\,dy'.$$

But we have (Art. 219 Cal.) for the length of the curve

$$w = -2\sqrt{2r(2r - y)}$$

reckoning from C; and substituting $2r - y'$ for y we have $w = -2\sqrt{2ry'}$, which gives, for the distance of the center of gravity from the point C, $\frac{\int y'dw}{w} = \frac{\sqrt{2r}\int y'^{1/2}\,dy'}{-2\sqrt{2ry'}}$

$$= \frac{1}{-2\sqrt{y'}}\int y'^{1/2}\,dy' = \frac{^2/_3\, y'^{\,3/2}}{2y'^{\,1/2}} = \tfrac{1}{3}\,y'.$$

Making $y' = 2r$ we find the center of gravity of a cycloid to be in its axis, and at a distance from its upper extremity, equal to one-third of the diameter of the generating circle.

To find the center of gravity of a plane surface.

Let the surface be bounded by the curve B A C and the double ordinate B D C. and be symmetrical about the line A D. This line will then contain the center of gravity of the surface, and the question will be, to determine its distance from the plane of reference A Y, perpendicular to A D. Since the center of gravity of each half of the surface is at the same distance from the plane of reference as that of the whole, we will take the upper half, A B D, for our purpose. Suppose the surface to have been generated by the movement of the ordinate commencing at A, and having its extremities always in the bounding line or curve and its middle point in the axis A D, until it arrives at B C. Let D F represent the rate at which the axis A D was increasing (or dx) when the generating ordinate arrived at B D then the rectangle D B E F will represent the rate at which the surface was generated at the same moment, for it is the surface that *would have been* generated, if the ordinate had continued to move, unchanged, at a uniform rate for the same unit of time as that required by D F, and this rate will be expressed by B D, D F or ydx.

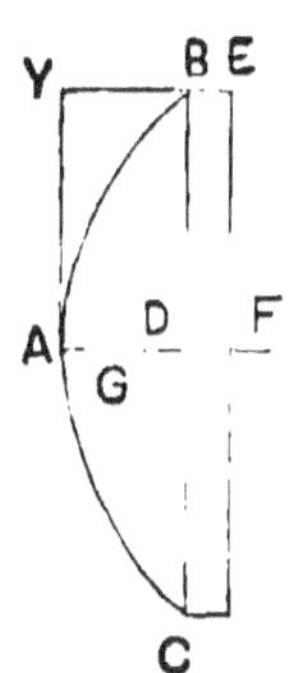

Since this symbol represents that which exists at the ordinate B D, its moment will be found by multiplying it by the distance of B D from the plane of reference, that is, B D, D F, B Y or $xydx$. But the rectangle represents the rate of increase of the sum of all the parts of the surface, and therefore, the product of the rectangle by its distance from the plane of reference will represent the rate of increase of the sum of the products of all these parts into their respective distances from the same plane; that is, the rate of increase (or differential) of the sum of the moments. Hence, $\int xydx$ will represent the sum of the moments, and this divided by the surface will give the distance of its center of gravity from the plane of reference.

Example.

To find the center of gravity of a surface bounded by a parabola and its double ordinate.

Let the plane of reference be tangent to the parabola at its vertex. In this case $xy\,dx$ becomes $\sqrt{2p}x^{3/2}\,dx$ and the surface is $\frac{2}{3}\,xy = \frac{2}{3}\,\sqrt{2p}\,x^{3/2}$ and hence for $\frac{\int xydx}{s}$ we shall have

$$\frac{\int\sqrt{2p}\;x^{3/2}\,dx}{\frac{2}{3}\sqrt{2p}\;x^{3/2}} = \frac{\int x^{3/2}\,dx}{\frac{2}{3}\,x^{3/2}} = \frac{{}^{2}/_{5}\,x^{5/2}}{\frac{2}{3}\,x^{3/2}} = {}^{3}/_{5}\,x + \mathrm{C}.$$

But when $x = 0$ we have $s = 0$ and hence $\mathrm{C} = 0$ and the distance of the center of gravity of a segment of a parabola from its vertex is ${}^{3}/_{5}$ the length of its axis.

To find the center of gravity of a surface of revolution.

Suppose the surface to be generated by the flowing of the circumference of a circle, constantly parallel to itself, and constantly intersecting the meridian section of the surface, and whose center shall be always in the axis of revolution. Let C A F represent the meridian section of the surface; A B, its axis of revolution; A Y, the plane of reference; C F, the diameter of the generating circumference; O, the length of the meridian curve: and S. the area of the surface.

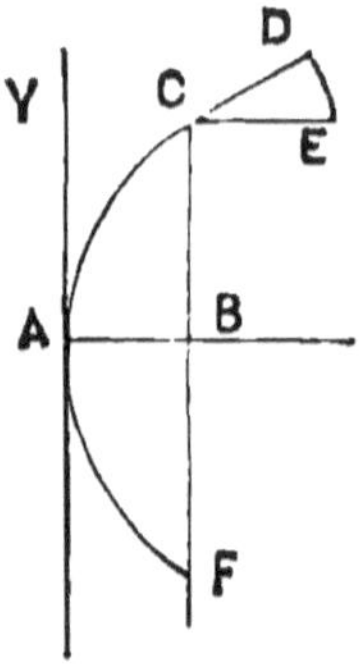

Suppose the center of the generating circumference to have arrived at the point B in its movement away from the plane of reference. At that instant every point in this circumference will have generated a meridian curve, and will tend to move in the direction of the tangent to the one in which it lies, and at a rate equal to that at which the curve was generated. Thus the point C will tend to move in the direction C D, tangent to the curve at C. Let C D represent the rate at which the point C is moving when it has arrived at that position; then the rate of movement of every point in the generating circumference will be the same; and the rate of generation of the surface will be represented by the surface of a cylinder whose base is the generating circumference, and whose height will be equal to C D (or the differential of O) which represents the rate at which every one of its points moves. This figure will be a cylinder because the surface which represents the rate, at that moment, must be that which *would* be generated at a uniform equal rate, which requires that the generating circumference should not change its dimensions. Hence if we make C E, parallel to the axis, = C D, the rate of generation of the surface of revolution at the moment the center of the generating circumference arrives at B will be represented by the surface of a cylinder of which C F is the diameter of the base and C E the altitude.

Now the surface of this cylinder being the symbol of the rate of increase of the sum of all the particles of the surface, will be represented by ds; and if we multiply it by the distance of C from the plane of reference, the product (xds) will be the symbol of the rate of increase of the sum of all the products arising from multiplying each particle into its distance from the plane of reference — that is, the moment of the rate of increase of the sum of the particles is the rate of increase of the sum of the moments of the particles, which will be represented by

$$xds = x.\, 2\pi y \sqrt{dx^2 + dy^2} = 2\pi xy\, (dx^2 + dy^2\,)^{1/2}$$

and the integral of this divided by the surface will give the distance of its center of gravity from the plane of reference.

EXAMPLE.

To find the center of gravity of the surface of a spherical segment.

Let A B be the axis of the segment perpendicular to its base and A Y the plane of reference; when the center of gravity will be in A B, and since $y^2 = r^2 - x^2$ the differential of the curve (dO) will be equal to

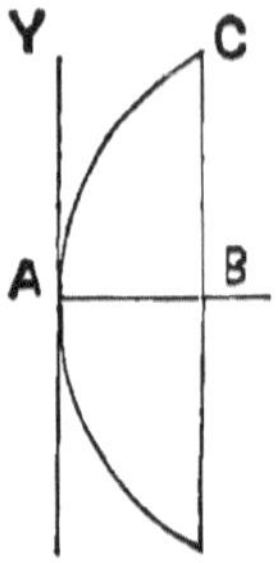

Y C A B

$$\sqrt{\frac{x^2\,dx^2}{r^2 - x^2} + dx^2} = dx\sqrt{\frac{r^2}{r^2 - x^2}}$$

and the differential of the surface (ds) will be

$$2\pi y\sqrt{\frac{r^2}{r^2 - x^2}}\,.\,dx = 2\pi r dx$$

while the surface itself is equal to

$$2\,\pi\,r\,x$$

hence for the distance of the center of gravity from the plane of reference $\left(\frac{\int x\,d\,s}{s}\right)$ we have

$$\frac{\int 2\pi r x dx}{2\pi r x} = \frac{\int x dx}{x} = \frac{x}{2}$$

or the center of gravity of the surface of a spherical segment is at a distance from its vertex equal to half its height; and making $x = \mathrm{R}$, the center of gravity of the surface of a hemisphere is on the middle of the radius perpendicular to its base.

To find the center of gravity of a solid of revolution.

Let C A B represent the meridian curve of the solid. A F its axis. A Y the plane of reference. G the center of gravity in the axis A B. Call the solid V, and suppose it to be generated by a circle flowing constantly parallel to itself from the vertex A, and varying its diameter in such a way that its circumference shall always intersect the meridian curve on the surface of the solid. Then if D F represent the rate of movement of the center of the circle (or dx) when it is at D, the cylinder having B D for the radius of its base and D F for its altitude, will represent the rate of generation of the solid at the same moment, and will be equal to $\pi y^2 dx$. Multiplying this by x, we have the moment of the rate of increase, which as we have seen, is the rate of increase of the sum of the moments of all the particles in the solid; and the integral of this is the sum of the same moments, and equal to the moment of the entire body, or $v \times \mathrm{AG}$; hence

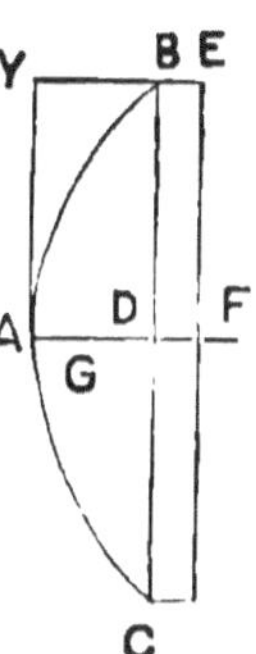

$$\mathrm{AG} = \frac{\int \pi y^2 x dx}{v}.$$

EXAMPLE.

To find the center of gravity of a hemisphere with the origin at the center and the plane of reference passing through the origin.

The volume of the spherical segment next to the center is

$$\pi\left(\mathrm{R}^2 x - \frac{x^3}{3}\right)$$

and we have for $\int \pi y^2 x dx$

whence

$$AG = \frac{\int \pi y^2 x dx}{v} = \frac{\pi \left(R^2 \frac{x^2}{2} - \frac{x^4}{4}\right)}{\pi \left(R^2 x - \frac{x^3}{3}\right)}$$

which becomes, by making $x = R$

$$AG = \frac{\frac{R^4}{4}}{R^3 - \frac{1}{3} R^3} = \frac{3}{8} R.$$

or the center of gravity of a hemisphere is on the radius perpendicular to its base and three-eighths of its length from the center of the sphere.

CENTER OF OSCILLATION.

A simple pendulum is a particle of matter suspended by an imponderable rod to a point about which it can vibrate freely.

A compound pendulum is any mass of matter suspended on a horizontal axis which does not pass through its center of gravity, and about which it can vibrate freely.

The center of oscillation of a pendulum is that point in which, if the entire mass be concentrated, the pendulum would vibrate in the same time as it would in its natural state. Hence the distance of the center of oscillation from the axis of suspension is the length of a simple pendulum vibrating in an equal time; and since the compound pendulum vibrates in the same time as the simple one, the force of gravity must act on each at the same distance from the axis of suspension; that is the resultant of all the forces of gravity, acting upon the particles of a compound pendulum, must pass through a point at the same distance from the axis of suspension, as the particle which composes a simple pendulum vibrating in the same time. Hence to find the center of oscillation we must find through what point this resultant passes.

Let S A be a pendulum composed of simple particles arranged in a straight line and suspended at the point S. Let *mp* represent the direct action of the force of gravity or weight of the particle *p;* it may be resolved into two component parts, viz. *np*, acting in the direction of the length of the pendulum, and *pk*, acting in a direction perpendicular to its length and which causes it to vibrate. This component, *pk* of the force of gravity will be the same for every particle in the pendulum, and would cause it to move so as to be constantly parallel to itself; but its action will be resisted by the axis of suspension, and thus cause the pendulum to turn around that axis; and the difference between the component of gravity *pk* and the resistance at S will be for each particle the effective force arising from its weight which tends to produce the vibration, and which increases with its distance from S. Hence the effective force of any particle, as *p* to cause a vibration may be represented by the weight of that particle into its distance from the point of suspension. This is called the moment of force of the particle with reference to the axis at S. Since, at every instant, these moments of force are acting in a direction perpendicular to the line of the pendulum, they will be parallel to each other, and the distance from their resultant to S will be found by multiplying every one into its distance from S, and dividing the sum of the products by the sum of the moments. The product of a moment of force into its distance from S is called, the moment of inertia of the particle with reference to the axis S, which is therefore equal to the weight of the particle into the square of its distance from the axis, and if *o* represent the center of oscillation we shall have

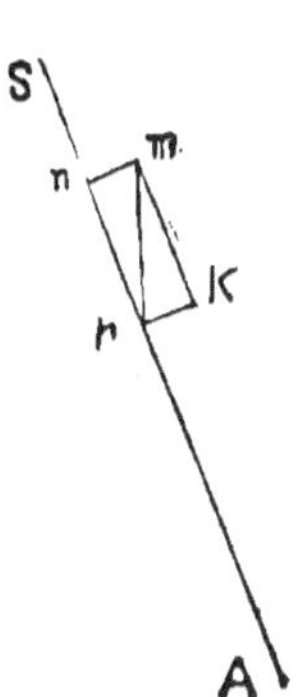

$$S\,o = \frac{\text{sum of moments of inertia}}{\text{sum of moments of force}},$$

which is equal to the sum of the products arising from multiplying each particle of matter into the square of its distance from the axis of suspension and dividing the sum of the products by the mass multiplied by the distance of the center of gravity from the same point.

It is of course impossible to obtain the sum of these products directly, since the particles of matter, although real are too minute to be estimated singly. We must therefore proceed indirectly by obtaining the rate at which these products increase, compared with that at which the distance of the center of oscillation increases, supposing the pendulum to be in a state of growth or increase.

Let S A be a straight inflexible pendulum composed of a single line of particles of matter, and suspended at the point S. Since the weight of any part of this pendulum will be in proportion to its length, and *vice versa*, the one may be represented by the other, and the center of gravity will be at the middle point. Let that point be C, and let O be the center of oscillation; while the weight of the pendulum is represented by w. Suppose now, the pendulum to be generated by growing from the point S downwards until it has reached the extremity A; and when at that point, let A B represent the part that would be generated in a unit of time at the rate then existing uniformly continued; then A B will represent the rate of increase of the pendulum at the point A— that is to say, the differential of w; and since the weight and length are proportional, A B multiplied by $\overline{SA}^2$ (or w^2dw) will represent the rate of increase of the sum of the products arising from multiplying each particle into the square of its distance from the axis of suspension; and the integral of this, or $\int w^2dw$ will be the sum itself. Dividing this sum by

S

C

O

A

B

the weight of the entire mass into the distance of its center of gravity from the axis of suspension — that is, by $w \, {}^{1}\!/_{2} \, w$, or ${}^{1}\!/_{2} \, w^2$ we have the distance of its center of oscillation from the axis; that is, $S\,O = \frac{\int w^2 dw}{{}^{1}\!/_{2} \, w^2} = \frac{2}{3} w$. Hence the center of oscillation of a straight pendulum of uniform weight, and without thickness, is at a distance from its point of suspension, equal to two-thirds of its length.

FORCE, TIME AND VELOCITY.

Force is that which tends to produce or destroy motion. Momentum is sometimes considered as a force, since when it exists in one body, it can impart motion to another by impact. But in such a case there is no motion *produced*— that is to say—the *quantity* of motion is no greater after impact than before; it is merely divided into parts: hence momentum is not properly a *producer* of motion, and therefore is not a *force*.

The quantity of motion or momentum of a body is a complex idea, of which the elements are, its mass or quantity of matter, and its velocity or rate of motion. For a given mass the momentum is in proportion to the velocity, and for a given velocity it is as the mass: hence the multiplying of either, multiplies the momentum; and if they both vary, we must, to express the idea, represent the momentum in the form of the product of two factors. There cannot of course be a real product of two such quantities, considered abstractly, any more than we can multiply an hour by a pound; nor can any single product represent an abstract momentum; we can only put them into the *form of a product* retaining both the factors. So if m represent the mass and v the velocity, the momentum will be represented by mv. Uniform velocity is, also, a complex idea, being the relation between the space passed over at a uniform rate by a moving body, and the time occupied by its passage. For a given time the velocity will be directly as the space, and for a given space the velocity will be inversely as the time. Hence the relation between the two will be represented in the form of a fraction thus, $\frac{S}{T}$ in which the numerator represents the space and the denominator the time. Here, again, it must not be supposed that there can be a real division of the space by the time; the quantities being wholly different in kind, one can no more be divided by the other than a mile can be divided by a pound. Nor, indeed, can the *absolute* value of the velocity be expressed by a single quotient if it could be found, since both quantities must be expressed or understood in order to express the relation between them.

To give an exact and simple measure of velocity, we use some unit of time, such as one second, one hour, one day or some other definite period, and the space indicating the velocity will be that which is passed over during that unit of time; but the space must be associated with the time or it would be meaningless. Hence, in the expression $\frac{S}{T}$, where S represents any space whatever and T any time whatever, we divide them both by the number of units in the time, which will give the space passed over in one unit of time and thus express the velocity by a single term. If, for instance, the space is five hundred feet and the time twenty seconds, we divide by twenty and the result will be twenty-five feet in the numerator and one second in the denominator, and the value of the velocity thus expressed is twenty-five feet in one second.

Now the differential of any quantity is represented by the change that *would* take place in a unit of time at the existing rate uniformly continued for that time; hence in this case the numerator of the fraction is the differential of the space, for it is the space that would be uniformly passed over in one unit of time, and the denominator is the differential of the time since it is the time that elapses during that same unit, and therefore the velocity will be expressed by $\frac{dS}{dT}$. Since the rate at which time passes never varies, dT will be a constant quantity, and if the velocity is constant then dS

will also be constant, and the expression $\frac{d\,S}{d\,T}$ may be replaced by $\frac{S}{T}$ for the increments of the space and time will always be proportional to their rates of increase. But if the velocity is variable then it can be expressed, for any instant, only by $\frac{d\,S}{d\,T}$ in which case $d\,S$ *does not* indicate the space *actually* passed over in one unit of time, but the space that *would be* passed over at the existing rate continued for that unit. It will therefore itself be a variable; and the velocity at any moment therefore, instead of being represented by the space divided by the time, will be represented by the rate of change in the space divided by the rate of change in the time.

A constant force free to act will always be measured by its effect; that is, by the momentum produced in a unit of time, and if it continues to act it will produce an equal amount of momentum in each successive unit of time; and hence at the end of T units of time the entire momentum will be represented by the force and time combined in the form of a product; as $f\,T$. This method of representing momentum, like the former, is to show how the two ideas of force and time are combined, and does not mean that there can be a real product any more than before: the momentum is still a complex idea, but the elements are different from the former ones.

The velocity produced by a continuous force will of course be a variable one, but at any one instant the momentum will be represented by mv, v being the velocity at that instant, and as we have just seen will also be represented by $f\,T$,— we have $mv = f\,T$ or $v = \frac{f\,T}{m}$ or $f = \frac{mv}{T}$, in which the numerator represents the momentum produced, and the denominator the time occupied in producing it. If we divide both the numerator and the denominator by the number of units in the time, we shall have the momentum produced in one unit of time for the numerator, and the unit of time itself for the denominator; and these will represent the rate of change in each, and hence $f = \frac{m\,dv}{d\,\mathrm{T}}$ (the force being a uniform one, and m being constant) will be the same as $f = \frac{m\,v}{\mathrm{T}}$, since the increments being made at a uniform rate are always proportional to the rates themselves.

If however the force is variable its value can be expressed only by $\frac{m\,dv}{d\,\mathrm{T}}$ and that but at one point of time, for $m\,dv$ would not indicate the momentum actually acquired in one unit of time, but that which *would be* acquired if the force were to continue uniform for that unit of time, from the instant at which the value of the force is required. If then we represent a variable force by F we shall always have $\mathrm{F} = \frac{m\,dv}{d\,\mathrm{T}}$ or, since $dv = \frac{d\,\mathrm{S}}{d\,\mathrm{T}}$ we have $\mathrm{F} = \frac{md^2\,\mathrm{S}}{d\,\mathrm{T}^2}$.

If the force is that of gravity then the mass (or m) may be omitted, as it is always proportional to the force, and since that force is practically constant near the surface of the earth, we have in that case (calling the force g) $g = \frac{v}{\mathrm{T}}$ or $v = g\,\mathrm{T}$. The force of gravity being variable at a distance from the earth, if we represent it by G we shall have $\mathrm{G} = \frac{d^2\mathrm{S}}{d\,\mathrm{T}^2} = \frac{dv}{d\,\mathrm{T}}$; and since $v = \frac{d\,\mathrm{S}}{d\,\mathrm{T}}$ we have $\frac{\mathrm{G}}{v} = \frac{dv.\,d\,\mathrm{T}}{d\mathrm{S}.\,d\mathrm{T}} = \frac{dv}{d\,\mathrm{S}}$ hence $\mathrm{G}d\,\mathrm{S} = vdv$ and $\mathrm{G} = \frac{dv.\,v}{d\,\mathrm{S}}$.

To find the time, space and velocity near the earth's surface.

Since $v = g\,\mathrm{T}$. and also $v = \frac{d\,\mathrm{S}}{d\,\mathrm{T}}$ we have $g\,\mathrm{T} = \frac{d\,\mathrm{S}}{d\,\mathrm{T}}$ or $d\,\mathrm{S} = g\,\mathrm{T}\,d\,\mathrm{T}$ which becomes by integration $\mathrm{S} = g\,\frac{\mathrm{T}^2}{2}$.

From these equations we can determine any two of these quantities, having the third given.

To find the velocity of a body falling through a great distance.

Let r be the radius of the earth, g the force of gravity at its surface, G the force at a distance x from the center of the earth, where the body is supposed to be after falling t seconds, and a its distance when the fall commences. Then since the force of gravity varies inversely as the square of the distance from the center of the earth we shall have $G : g :: \frac{1}{x^2} : \frac{1}{r^2}$, wherefore $G = \frac{gr^2}{x^2}$. But we have found $G = \frac{vdv}{dS}$ and equating these values of G we have

$$\frac{vdv}{dS} = \frac{gr^2}{x^2}.$$

Since $S = a - x$ we have $dS = -dx$ and by substitution

$$vdv = -\frac{gr^2\,dx}{x^2} = -gr^2x^{-2}\,dx$$

which becomes by integration

$$\frac{v^2}{2} = gr^2x^{-1} \text{ or } v^2 = \frac{2gr^2}{x} + C.$$

when $v = 0$ we have $C = -\frac{2gr^2}{a}$ and therefore the velocity acquired by a body falling from rest will be

$$v^2 = \frac{2gr^2}{x} - \frac{2gr^2}{a} = \frac{2gr^2(a-x)}{ax}.$$

If $x = r$ and the body has arrived at the surface of the earth, we have

$$v = \sqrt{\frac{2gr(a-r)}{a}}.$$

If the body should fall from the moon, we should have $r = 3956$ miles, $a = 60r$ and $g = 32\frac{1}{6}$ feet (that being the velocity produced by the force of gravity in one second of time); and for the value of v

$$v = \sqrt{\frac{2g.r\,59r}{60r}} = \sqrt{\frac{59r.32\frac{1}{6}ft.}{30}} = 6.88 \text{ miles nearly}$$

r being taken 3956 miles.

If we consider the distance a as infinite, it does not follow that the velocity of a body falling to the earth through an infinite distance will be infinite; for then we shall have

$$v^2 = 2gr \text{ and } v = \sqrt{2gr} = 6.97 \text{ miles}$$

nearly; so that however far a body may fall, on arriving at the surface of the earth, it can never attain a velocity greater than 6.97 miles in one second of time.

To find the force of attraction excited by a line upon a point without it, in a direction perpendicular to the line.

Let A B be the line and P the point attracted by it. Let P A be perpendicular to A B and equal to a. Suppose the line to be generated by a point flowing from A toward B, and suppose that on arriving at B it is moving at a rate that would carry it to C in a unit of time; then B C will represent the rate of increase of the line, or dx. We must now enquire what will be the corresponding rate of increase of the attraction of the line for the point P in the required direction. The attraction of B C for the point P will be proportional to the mass of B C divided by $\overline{PB}^2$; for B C is not an integral part of the line A B, but a symbol representing a value which exists at the point B only.

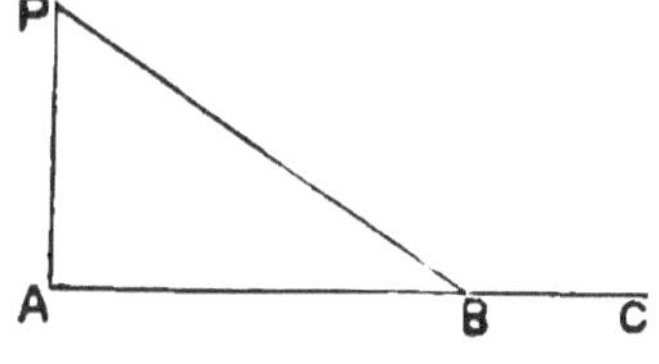

Let m represent the mass of one unit of the line B C supposed to exist at B, and m' that of P; and let k represent the attraction of one particle of each mass on one particle of the other; then the attraction of one unit of B C for P will be equal to $\frac{kmm'}{\overline{PB}^2}$; and since the rate of increase of this attraction is proportional to the rate of increase of the line itself, the attraction of the symbol B C (which represents this rate, or dx) for the point P will be equal to $\frac{kmm'dx}{\overline{PB}^2}$. Now this expresses the attraction of B C for P in the direction P B, and is to its attraction in the direction P A as P B : P A : : $\sqrt{a^2 + x^2} : a$. Hence the attraction of B C for P in the direction P A is equal to its attraction toward B multiplied by a and divided by $\sqrt{a^2 + x^2}$.

But the attraction of B C is the rate of increase of the attraction of the line, and is equal to $\frac{kmm'dx}{a^2 + x^2} \times \frac{a}{\sqrt{a^2 + x^2}}$ $= \frac{kmm'adx}{(a^2 + x^2)^{3/2}}$ which becomes, by integration, $\frac{kmm'x}{a\sqrt{a^2 + x^2}}$ (C being zero) which is the whole attraction of A B for P in the direction P A supposing $x =$ A B.

www.ingramcontent.com/pod-product-compliance
Lightning Source LLC
Chambersburg PA
CBHW022122290426
44112CB00008B/777

* 9 7 8 3 7 4 4 7 2 9 4 1 3 *